Cases on Engineering Management Education in Practice

Despo Ktoridou
University of Nicosia, Cyprus

A volume in the Advances in Higher Education and Professional Development (AHEPD) Book Series

IGI Global
PUBLISHER of TIMELY KNOWLEDGE

Published in the United States of America by
IGI Global
Business Science Reference (an imprint of IGI Global)
701 E. Chocolate Avenue
Hershey PA, USA 17033
Tel: 717-533-8845
Fax: 717-533-8661
E-mail: cust@igi-global.com
Web site: http://www.igi-global.com

Library of Congress Cataloging-in-Publication Data

Names: Ktoridou, Despo, 1968- editor.
Title: Cases on engineering management education in practice / edited by Despo Ktoridou.
Description: Hershey, PA : Engineering Science Reference, 2021. | Includes bibliographical references and index. | Summary: "This book explores curriculum development, instructional design, and pedagogies of engineering management learning initiatives"-- Provided by publisher.
Identifiers: LCCN 2020006881 (print) | LCCN 2020006882 (ebook) | ISBN 9781799840633 (hardcover) | ISBN 9781799854982 (paperback) | ISBN 9781799840640 (ebook)
Subjects: LCSH: Engineering--Management--Study and teaching (Higher)
Classification: LCC TA190 .C38 2021 (print) | LCC TA190 (ebook) | DDC 620.0071/1--dc23
LC record available at https://lccn.loc.gov/2020006881
LC ebook record available at https://lccn.loc.gov/2020006882

This book is published in the IGI Global book series Advances in Higher Education and Professional Development (AHEPD) (ISSN: 2327-6983; eISSN: 2327-6991)

British Cataloguing in Publication Data
A Cataloguing in Publication record for this book is available from the British Library.

Advances in Higher Education and Professional Development (AHEPD) Book Series

ISSN:2327-6983
EISSN:2327-6991

Editor-in-Chief: Jared Keengwe, University of North Dakota, USA

MISSION

As world economies continue to shift and change in response to global financial situations, job markets have begun to demand a more highly-skilled workforce. In many industries a college degree is the minimum requirement and further educational development is expected to advance. With these current trends in mind, the **Advances in Higher Education & Professional Development (AHEPD) Book Series** provides an outlet for researchers and academics to publish their research in these areas and to distribute these works to practitioners and other researchers.

AHEPD encompasses all research dealing with higher education pedagogy, development, and curriculum design, as well as all areas of professional development, regardless of focus.

COVERAGE

- Adult Education
- Assessment in Higher Education
- Career Training
- Coaching and Mentoring
- Continuing Professional Development
- Governance in Higher Education
- Higher Education Policy
- Pedagogy of Teaching Higher Education
- Vocational Education

IGI Global is currently accepting manuscripts for publication within this series. To submit a proposal for a volume in this series, please contact our Acquisition Editors at Acquisitions@igi-global.com or visit: http://www.igi-global.com/publish/.

The Advances in Higher Education and Professional Development (AHEPD) Book Series (ISSN 2327-6983) is published by IGI Global, 701 E. Chocolate Avenue, Hershey, PA 17033-1240, USA, www.igi-global.com. This series is composed of titles available for purchase individually; each title is edited to be contextually exclusive from any other title within the series. For pricing and ordering information please visit http://www.igi-global.com/book-series/advances-higher-education-professional-development/73681. Postmaster: Send all address changes to above address.

Titles in this Series

For a list of additional titles in this series, please visit:
http://www.igi-global.com/book-series/advances-higher-education-professional-development/73681

University-Industry Collaboration Strategies in the Digital Era
Durmuş Günay (Maltepe University, Turkey) Tuncer Asunakutlu (İzmir Bakircay University, Turkey) and Orkun Yildiz (Izmir Democracy University, Turkey)
Information Science Reference • © 2021 • 300pp • H/C (ISBN: 9781799839019) • US $195.00

Community Colleges and Workforce Preparation in the 21st Century
Dimitra Smith (Texas A&M University at Commerce, USA) Rhonda Lewis (Collin College, USA) and Leslie Singleton (Sam Houston State University, USA)
Information Science Reference • © 2021 • 335pp • H/C (ISBN: 9781799841234) • US $175.00

Developing E-Assessment Tools and Technologies in Higher Education
Piero Espino-Román (Universidad Politécnica de Sinaloa, Mexico) Eugenia Olaguez-Torres (Universidad Politécnica de Sinaloa, Mexico) and Luz Marlene Inzunza-Moraila (Universidad Politécnica de Sinaloa, Mexico)
Information Science Reference • © 2021 • 300pp • H/C (ISBN: 9781799849872) • US $185.00

Evidence-Based Tools for Improving Student Retention and Success in Higher Education
Jeffrey Herron (Campbellsville University, USA)
Information Science Reference • © 2021 • 280pp • H/C (ISBN: 9781799829188) • US $195.00

Navigating Post-Doctoral Career Placement, Research, and Professionalism
Noran L. Moffett (Fayetteville State University, USA)
Information Science Reference • © 2021 • 300pp • H/C (ISBN: 9781799850656) • US $195.00

For an entire list of titles in this series, please visit:
http://www.igi-global.com/book-series/advances-higher-education-professional-development/73681

IGI Global
PUBLISHER of TIMELY KNOWLEDGE

701 East Chocolate Avenue, Hershey, PA 17033, USA
Tel: 717-533-8845 x100 • Fax: 717-533-8661
E-Mail: cust@igi-global.com • www.igi-global.com

Editorial Advisory Board

Table of Contents

Detailed Table of Contents

Training in communication skills is considered extremely important in the engineering profession. However, educational organisations and most specifically engineering programs and departments have often been criticised for failing to adequately prepare engineering students for the situations they will face in the workplace. This chapter describes a technical writing and research course that is offered as a required course to engineering students and analyses the advantages and limitations of the course pointing to changes in the course development that will enable students to perform successfully as communicators in the workplace.

This chapter focuses on the design and redesign of an online engineering management course, based on a socio-constructivist approach. At first, the theoretical and contextual premises will be presented with a focus on the suggested teaching and learning methods to acquire domain-related knowledge and crucial skills and on the importance of learning scenario to support an effective learning design. After the background introduction, a user case will be described, focusing on the course online environment and its tools, on the proposed pedagogical strategies and above all, on how instructors can obtain and analyze useful educational data from various sources. Finally, some redesign recommendations will be provided to better use educational data for continuous course improvement.

In the dynamic world that is associated with unpredictable informatization and globalization, it is indispensable for education to be sustainable and continuously

Mixed methods were used to achieve the above. Initially, two focus groups were conducted, and at a later stage, 126 engineers reported through a questionnaire the most common challenges in the transition from engineering to management and then a few were also interviewed. Results demonstrate that skills such as delegation, communication, convincing, coaching, and guiding others pose important challenges. In relation to the second inquiry, it is proposed that, other than management theory, offering examples, opportunity to practice with feedback, and case-based learning can help minimise the challenges. Also, leadership skills, such as delegation, developing personality, cross-cultural understanding, and managing diversity, can be strategically used to facilitate learning in the field of engineering and better prepare engineers in their transition to management.

The aim of this chapter is to bridge the gap between literature about leadership studies and engineering education to develop future engineering leaders. The chapter focuses upon the hindrances that are faced in integrating leadership programs in the engineering circular. To this end, the chapter employed qualitative methodology under which it reviewed literature on multiple lines such as identifying significance of leadership for engineering students, identification of leadership practices and skills taught to engineering students, and the way this can be improved. Key findings of chapter emphasized on leadership as an important skill for future engineers thus demanding need for equipping engineers with skills needed to become an effective leader. It is found that having engineers to develop leadership skills will enable them to perform their role effectively, effectively allocate and utilize available resources. Thus, the chapter recommends integration of leadership knowledge and skills in engineering curricula through usage of variety of tools such as experiential learning and self-awareness.

The chapter aims at addressing some of the critical questions that will be of immense help to engineering graduates who are likely to consider pursuing management degrees. Data were obtained from a sample of 96 techno managers to explore their perception of career growth and opportunities after pursuing their management education. In particular, the chapter examines in detail the relationship between management education and career satisfaction in techno managers. Also, factors that influence in choosing a management degree were studied. A few of the critical findings from the study were that gender has not shown any significant difference in satisfaction. At the same time, experience has shown a significant difference in satisfaction. From the results obtained, it could be concluded that the techno managers were found to be satisfied with the career choice as an MBA after engineering.

Foreword

Engineering management is the application of the practice of management to engineering. Training for engineering management presupposes that the individual has completed an undergraduate degree specializing in one or more areas of fundamental engineering or technology innovation, such as mechanical, materials, biomedical, telecommunications, civil, mining, systems, industrial, requirements, design engineering, product or process engineering (e.g. supply chain) and has then opted to study the combined application of engineering within a management context to address complex sociotechnical systems and commensurate responses in the social-cyber-physical world. Some of these new initiatives may be greenfield implementations while others have existing infrastructure that requires engineering management, such as in brownfield implementations toward retrofitting, upgrade, or extension to mention but a few configurations.

It is not enough for an individual to be equipped with the latest technical knowledge on "how something works". Equally, if not more important, is the ability to organize and mobilise individual componentry (i.e. assets), human factors toward administration and specialisation, design and construction and the commensurate processes required to make a project, a blueprint or architecture, a reality. Increasingly important is an understanding of non-technical issues that I like to term "human factors in engineering" that refer to those things pertaining to society and the regulatory environment. These two areas, can bring an engineering endeavor to a standstill if not considered long before the actual implementation phased approach. Right from the outset, engineering managers and engineers themselves, have to consider how to embed ethics into the design process, consider consultation with citizenry (e.g. the enactment of a new 5G base station near a School), conduct a business case, and become aware of the environmental conditions in which they are conducting business.

Failure to consider human factors in engineering can be detrimental to any project and lead to a delayed "go live" date, a product that does not achieve desired functionality, a significantly overbudget project, or at the worst case scenario an incomplete large-scale project that stalls and must go into receivership. This is why Despo Ktoridou's edited volume incorporating invaluable perspectives from global authors in relation to engineering management education is so vital. As complexity grows in systems of all kinds, thanks to computational abilities in processing and storage and AI-based algorithms, additive manufacturing and 3D printing techniques, the adoption and application of new materials, even financial markets, we must ensure our students are ready for the complex value chains that have pervaded, the adaptive capabilities that are required in a turbulent environment that can see volatile fluctuations in the market for factors that would never before have been considered relevant.

In short, how do we impart this knowledge to our predominantly Masters level students, most of whom have had a taste of industry and are aware of some of the technical challenges facing engineering management at large? How do we emphasise cross-disciplinarity? Is it by forming unique groups to tackle large-scale case challenges through problem-based learning approaches? Is it to pose scenarios and come out with what-if events? It may even be going so far as to simulate an environment that may require the orderings of tens of thousands of smaller parts, consultation with legal experts in advising on changing laws, or otherwise.

This book stands as a wonderful contribution to coming forward in educating those who are responsible for engineering systems by emphasising the socio-technical-legal aspects as commensurately important in the curriculum as technical innovation management. It also lends educators the ability to reflect on their practice of engineering management and to try different things in their curriculum design that they may not have considered previously.

I'd like to take this opportunity to congratulate the editor and the contributing authors to this great volume.

Katina Michael
Arizona State University, USA

Katina Michael *is a professor at Arizona State University, holding a joint appointment in the School for the Future of Innovation in Society and School of Computing, Informatics and Decisions Systems Engineering. She is also the director of the Society Policy Engineering Collective (SPEC) and the Founding Editor-in-Chief of the IEEE Transactions on Technology and Society. Katina is a senior member of the IEEE and a Public Interest Technology advocate who studies the social implications of technology. She has held 13 annual workshops in the social implications of national security space and chaired 3 international symposia on technology and society (ISTAS) in Wollongong, Toronto and Phoenix. She is the Senior Editor of the socio-economic impact section in IEEE Consumer Electronics Magazine and was the editor in chief of the award-winning IEEE Technology and Society Magazine. In 2019 she took on the role of working group chair for the IEEE P2089 standard. In 2017, she received the Brian M. O'Connell SSIT Distinguished Service Award. In the 1990s, Katina was employed as a senior network planner at Nortel Networks and systems analyst at Andersen Consulting and OTIS.*

In 2002 she entered academia as a lecturer at the University of Wollongong and in 2013 became a member of the executive team in the Faculty of Engineering and Information Sciences as the Associate Dean – International, overseeing 8 partner and twinning arrangements for the University in the UAE, Malaysia, Singapore, China and Hong Kong. Katina completed a Masters degree in transnational crime prevention in the Faculty of Law with a UOW postgraduate scholarship in 2007, and was later invited to teach cybercrime in the law school. Her PhD was on systems of innovation in the automatic identification industry.

Preface

The continuous technological, economic and social challenges, especially since the onset of the third millennium, make it imperative for higher educational institutions to equip students with the necessary knowledge, skills and competences in order to seek employment in such a challenging, global context. One group of employees that can contribute in meeting today's challenges are engineers-managers. Engineering managers in a contemporary context are not simply about conceiving and designing engineering systems. Most importantly, they lead projects, drive research and development, create strategic plans, handle financing and evaluate market opportunities, among others.

An Engineering Management graduate program usually has a three dimensional foci: Technical people, projects and organizations. It prepares engineer and science graduates to become future industry leaders by building a solid foundation for a business to gain long-term competitive advantages. Such a programme, focuses on two distinct disciplines, engineering and management and particularly on innovation and technology management in specific engineering disciplines.

A key and rather pervasive question, which puzzles educational institutions and companies for several years, is how to facilitate and support the transition of engineers to managers. What kind of education or training is required to prepare engineers for a career in management? How can existing engineers manage their career as managers and leaders? Although these inquires have been central in engineering studies for decades, they are still prevalent and ever present. Therefore, the current book is a timely contribution to thinking systematically about a transition of engineers to positions of management and leadership.

In the following chapters, distinguished academics and industry leaders from five different countries, contribute eight real life scenarios to explore the transition of engineers to management. Owing to a wide-ranging research area, their work expands to include significant engineering management pedagogies, instructional design strategies, and career management through sustainable lenses. The objectives are to: a) formulate a better understanding of engineering management education practices b) explore how these practices adapt to the ever-changing technological, social and economic forces, and c) identify how we can best prepare future engineer business generalists for management positions.

Among the several chapter proposals received, the editor organized the book under two themes: a) Engineering Management Curriculum Development and Instructional Design, with four chapters and b) Engineering Management Transition and Careers, with four chapters. All eight chapters are easy-to-read, ready to be applied, offering real world or authentic practices exposing educators and practitioners to multiple sources of viewpoints on engineering management education practices, transition and careers.

More specifically, there are contributions by authors from Cyprus, Greece, Italy, UK and India, which enhances the value of this book. The eight chapters offer a glance on a wide range of issues in Engineering Management Education Practices: A Measurement Model for Collaborative Online Learning in Postgraduate Engineering Management Studies; Technical writing and Communication Skills ; Collaborative online Learning; The role of Learning Scenarios and Learning Analytics in the design and redesign of an online socio-constructivist Engineering Management courses; Innovative learning pedagogies; Emotional Intelligence in Engineering Management Education; Challenges on Engineering to Management Transition; The Power of Leadership in Engineering Management Education and Technology Management Education, Satisfaction and Career Growth.

As summary, the key ideas from each chapter are presented below:

In Chapter 1, Dr Leonidas Efthymiou, Dr Alex Zarifis and Ms Yianna Orphanidou introduce certain metrics (objectives, critical success factors, key performance indicators, and targets) along with a handful of constructivism techniques, namely, ‘peer interaction’, ‘forum activities’, ‘learning by doing’ and ‘systematic feedback’. Linking the constructs of social constructivism with quantification enables as to develop a rational model of performance measurement, serving as a navigation instrument for instructors, instructional designers and learners.

In Chapter 2, Dr Anna-Maria Andreou, Ministry of Education, Youth, Culture and Sport, Cyprus addresses the importance of training in communication skills in the engineering profession. Her chapter describes a technical writing and research course that is offered as a required course to all engineering students and at the and analyses the advantages and limitations of the course pointing to changes in the course development that will enable students to perform successfully as communicators in the workplace.

In Chapter 3, Dr Mary Grammatikou, Dr Nadia Sansone, Dr Dimitris Pantazatos, Dr Donatella Cesareni and Dr Vasilis Maglaris focus on the design and redesign of an online engineering management course based on a socio-constructivist approach. Their theoretical and contextual premises focus on the suggested teaching and learning methods to acquire domain-related knowledge and crucial skills and on the importance of learning scenarios to support an effective learning design. Following the background introduction, the authors describe a user case, focusing on the course online environment and its tools, on the proposed pedagogical strategies and, above all, on how instructors can obtain and analyze useful educational data from various sources. Finally, they provide some redesign recommendations to better use educational data for continuous course improvement.

In Chapter 4, Dr Marcos Komodromos discusses how innovative learning and education practices for modern pedagogy have been developed in the European universities' system. Specifically, the author provides a delineation of the attributes, pedagogy, ideology, ethical dimensions, and practical applications of innovative learning education practices. In the chapter, the author identified the different practices such as Massive Open Online Courses (MOOCs), blended learning, and flipped classrooms that have been applied at different European universities. Thus, the information discussed in this chapter is relevant to university educators and electronic learning practitioners.

In Chapter 5, Dr. Marilena Antoniadou, Dr. Mark Crowder and Mr. George Andreakos aim to increase understanding of how engineering students can benefit from integrating Emotional Intelligence (EI) into engineering curricula at universities. In particular, the chapter explores the nature of EI and argues for the greater use of EI within engineering education, but also considers the challenges of placing an emphasis on EI within this field. The chapter makes recommendations for how EI skills can be incorporated into engineering education and how universities can seize the opportunity to shape the modern engineer and advance the standing of engineering in the future. The chapter's contribution lies in raising awareness not just about the benefits of integrating EI within engineering education, but also on the challenges that an empathetic behavior entail. They argue that university education needs to be able to prepare graduates not just with engineering fundamentals, but also for success and actual on-the-job EI skills.

In Chapter 6, Dr Leonidas Efthymiou, Dr Epaminondas Epaminonda and Professor Despo Ktoridou focus on the challenges in the transition from engineering to management and discuss how management education may assist in this transition. Mixed methods were used to achieve the above. Initially, two focus groups were conducted and at a later stage 126 engineers reported through a questionnaire the most common challenges in the transition from engineering to management and then a few of them were also interviewed. Results demonstrate that skills such as delegation, communication, convincing, coaching and guiding others pose important challenges. In relation to the second inquiry, it is proposed that, other than management theory, offering examples, opportunity to practice with feedback and case-based learning can help minimize the challenges. Also, leadership skills, such as delegation, developing personality, cross-cultural understanding and managing diversity can be strategically used to facilitate learning in the field of engineering and better prepare engineers in their transition to management.

In Chapter 7 Dr. Zeyad Waleed Abualhamael, aims to identify the importance of leadership studies for engineering students; provide better and wide understanding of the adopted leadership and management studies within engineering education; expand our knowledge in adopting more progressive higher educational institutions to improve the education of engineering students as well as addressing leadership challenges in the engineering context. A fundamental significance of this chapter is to bridge the gap between leadership studies and engineering education to develop motivated, inspired, prepared and positioned engineers that can direct project teams and create a collective working environment. Eventually, the chapter addresses the incorporation of leadership development and engineering education programs as a response to global processional demands.

In Chapter 8, Dr Venugopal Merugu and Dr Rahul Thangeda address some of the critical questions that may assist engineering graduates who are likely to consider pursuing management degrees. Mainly, the chapter examines the relationship between management education and career satisfaction in techno managers, the perceptions of 96 techno managers on career growth and opportunities after pursuing their management education are explored. In addition, factors that influence the selection of a management degree are examined.

Significant findings are that gender has not shown any significant difference in satisfaction while experience has shown a significant difference in satisfaction. From the results obtained, the authors conclude that the techno managers were found to be satisfied choosing an MBA degree after engineering.

I hope this book would serve as a useful source for scholars, researchers, undergraduate and graduate students, as well as practitioners working in the field of Engineering Management as well as education in various disciplines, such as Management, Marketing, Finance, Leadership, Organizational Behavior, Human Resource Management, Statistics, Law, Sociology, Computer Science, and Information Technology. Furthermore, I hope the book will provide insights and support executives concerned with the management of engineering education practices.

Despo Ktoridou
University of Nicosia, Cyprus

Acknowledgment

First and foremost, I would like to thank God for giving me the strength and knowledge to undertake this Book project and complete it successfully.

There are a number of people I would like to thank for their contribution in getting this book published.

First, I would like to express my sincere gratitude to my colleagues Dr Epaminondas Epaminonda and Dr Leonidas Efthymiou for their unconditional help and support during this two-year journey. They helped whenever they were asked to do so, always acting consistently and professionally. Especially at the final stages of preparing the book their help had been instrumental for its successful completion. I thank them dearly.

My special Thanks go to all 20 authors coming from Cyprus, Greece, Italy, UK and India, who contributed with their time, knowledge and expertise for the successful completion of this book project.

My sincere acknowledgment also goes to the reviewers for their valuable contributions regarding the improvement of quality, coherence, and content presentation of the book's chapters.

Finally, I would like also to acknowledge the contribution of Dr Achilleas Karayiannis who embarked with me on this project but personal circumstances did not allow him to remain part of the team.

Despo Ktoridou
University of Nicosia, Cyprus

Section 1

Engineering Management Curriculum Development and Instructional Design

Chapter 1
A Measurement Model for Collaborative Online Learning in Postgraduate Engineering Management Studies

Leonidas Efthymiou
https://orcid.org/0000-0001-8099-5885
University of Nicosia, Cyprus

Alex Zarifis
https://orcid.org/0000-0003-3103-4601
Loughborough University, UK

Yianna Orphanidou
https://orcid.org/0000-0003-1933-4409
University of Nicosia, Cyprus

EXECUTIVE SUMMARY

Although the approach of social constructivism is not new (its origins are dated back to the pre-World War II era), it can be used along with novel learning strategies to facilitate quality online learning. The progress of technology, learning platforms and digital resources, together with certain social constructivism techniques, enable engineering practitioners to study in postgraduate management programs that replicate face-to-face environments. In this chapter, the authors introduce certain metrics (objectives, critical success factors, key performance indicators, and targets) along with a handful of constructivism techniques, namely, 'peer interaction', 'forum activities', 'learning by doing', and 'systematic feedback'. Linking the constructs of social constructivism with quantification enables us to develop a rational model of performance measurement, serving as a navigation instrument for instructors, instructional designers, and learners.

DOI: 10.4018/978-1-7998-4063-3.ch001

INTRODUCTION

This conceptual chapter brings together Social Constructivism and Performance Measurement (PM) to offer a measurement model for postgraduate online students in Engineering Management. Departing from the pedagogy of social constructivism, (Dewey, 1938), the chapter reviews a number of learning techniques, namely, 'social interactions', 'online communities', 'learning by doing', 'systematic feedback', 'continuous assessment', and the role of 'instructor as facilitator' – all of them intended to enhance fully interactive and engaging online learning experiences.

The social constructivist approach in online learning involves interaction with instructors and peers in forum activities. High levels of interaction among learners result to a positive attitude, greater satisfaction and therefore, successful distance education (Desai et al., 2009; Mooney, 2008). In addition, the environment is instructor-led where learners are guided through the content and learning activities at specified times (Rhode, 2009). Often, the instructor is a facilitator rather than a surrogate teacher or conveyor of information (Briggs, 2013, Smolin and Lawless, 2003). Instructors as well as peers systematically receive and offer feedback, which is an essential component of effective teaching and learning (Boling et al, 2011; Chan & Repman, 1999).

The use of social constructivism techniques along with a number of metrics and tools enable us to develop a student performance measurement model with Objectives, Critical Success Factors (CSFs), Key Performance Indicators (KPIs) and Targets. Linking the constructs of social constructivism with quantification is deemed important as it progresses both fields and offers a rational model to instructors and students. Hence, the questions this study seeks to address concern an attempt to quantify the online learning experience, namely: what metrics can we combine to measure the online learning experience? What do we want to measure and how can we best serve students' needs proactively? This conceptual chapter, then, is not a literature review. It is rather a navigation instrument; a scholar's guide of how to systematically prepare by quantifying online learning performance. This type of conceptual study is called a 'model' (Jaakkola, 2020) inasmuch as it seeks to build a theoretical framework that predicts relationships between concepts; identifies connections between constructs and proposes particular outcomes (Cornelissen 2017).

To our knowledge, this is the first time a full performance measurement model is designed for online post-graduate learners in engineering management whereas it is likely to be useful to instructors and instructional designers. The proposed model is intended specifically for formal, instructor-led, engaging, online learning experiences, aligned completely with the requirements of National Standards in Higher Education and Quality Assurance agencies.

The chapter is organized as follows: the next section is the Methodology, which explains the chapter's purpose, research design and methodological approach. Then, we review the literature on social constructivism prior to introducing measurement and metrics. This in turn enables us to present and discuss the proposed measurement model for strategic learning. We conclude with a summary of the implications of our study.

METHODOLOGY

The purpose of this chapter is to build a theoretical framework of how online learning performance can be broken down to measurable pieces; and how all the pieces together can facilitate quality online learning experiences. As mentioned in the introduction, this kind of conceptual study is called 'model' (Jaakkola, 2020) inasmuch as it seeks to build a framework that predicts relationships between concepts; identifies connections between constructs, and proposes particular outcomes (Cornelissen 2017; Huang & Rust 2018; Payne et al. 2017). Conceptual studies, like the current chapter, are important as they 'bridge existing theories in interesting ways, link work across disciplines, provide multi-level insights, and broaden the scope of our thinking" (Gilson & Goldberg 2015, p. 128). Towards this end, our analysis relies on a systematic review and synthesis of a) literature on social constructivism and b) performance measurement. These two fields are the 'conceptual ingredients' (Jaakkola, 2020: 2), which have been selected to develop a measurement model for quality online learning.

LITERATURE REVIEW

The origins of 'Distance education' can be traced back to the mid-19th century (Spector et al., 2008) when postal services and railway networks enabled an early form of course delivery via correspondence (Moore & Kearsley, 2011; Keegan, 1996). Gradually, distance learning evolved to include mail, cable television, satellite, videotapes and CDs (Whiteley, 2006). Then, computers and the Internet became involved in the delivery of learning and quickly, education entered a phase of time-space compression. 'Distance' was no longer an obstacle and not any more necessary to describe the limitations associated with education (Guilar & Loring, 2008, cited in Moore et al., 2011). The educational process entered the learner's space, pace and time (Anastasiades, 2002a). From this time forward, various names were assigned to distance education, such as eLearning (Tham & Werner, 2005) technology mediated learning, online collaborative learning, virtual learning and web-based learning (Conrad, 2006), all of them encompassing electronic and digital characteristics.

The types of distance education (Keegan, 1996) depend on the learning objectives, access (a spectrum from entirely virtual to blended with different levels of face to face engagement), tools and techniques used, the technology employed, type of content and the targeted learner market. Massive Open Online Courses (MOOCs), for example, which initially appeared in 2008 (Leontyev & Baranov, 2013), is a type of self-regulated learning, offered to massive numbers of learners in an open digital environment. It is free and available for everyone to enroll whereas learning outcomes are often abstract and not measurable (Lin et al, 2018). Learners have different learning goals, low peer interaction (often they may only know their peers' username), and no obligation to complete the course (Alraimi et al., 2015). Other models are characterized by the so-called airline budget analogy, where the base price might simply be delivery of material and anything additional, such as instructor feedback, comes at additional cost. Some programs still share the characteristics of self-study 'correspondence learning', inasmuch as students receive study guides, textbooks, assignments and other study materials through the post, but make use of technology to invigilate final exams remotely (IATA, 2018). Also, regardless of the model, online learning usually takes place on a specific web-based area, which is often identified as a digital platform; a Learning Management System (LMS); a Course Management System (CMS); a Virtual Learning Environment (VLE); a Virtual Classroom Setting (VCS); or an Online Learning Environment (OLE), to mention just a few (Spector at al, 2008).

While the types and models for online learning vary, the pedagogy presented in this chapter is located in the theoretical framework of social constructivism. Drawing on the work of major educational theorists, such as Dewey (1938), this chapter aligns with the social-constructivist view inasmuch as the process of learning and knowledge generation relies on participation in social interactions. The review

Figure 1. The five areas the literature review brings together

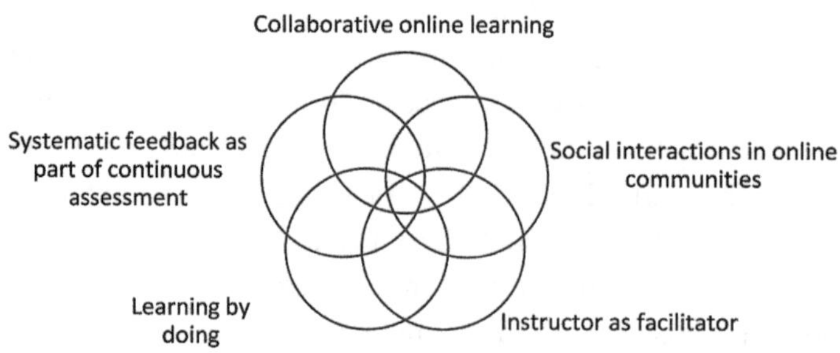

below refers to a number of learning strategies, as part of a social constructivism approach, starting with 'Social interactions in learning communities'.

Social Interactions in Online Communities

Unlike the objectivist approach, which only favours learner-teacher and learner-content interaction, social constructivism emphasizes on the natural ability of engineer students to think, gather information and generate knowledge through student-student interaction (Dewey, 1938). For many, interaction is one of the most important components in a learning experience (Vrasidas, 2000; Vygotsky, 1978) and central in distance education research (McIsaac & Gunawardena, 1996; Wagner, 1994). Thanks to the development of digital platforms in several fields (e.g. Efthymiou, 2018; Efthymiou, 2019; Batiz-Lazo et al, 2016), including learning platforms, students' social interaction is no longer restricted to physical classrooms and campuses.

Online platform interactions are a great opportunity to replicate traditional face-to-face classrooms and at the same time simulate a workplace environment for learners (in order to prepare them better for a career advancement). For example, a typical day at work for engineering managers includes frequent communication with colleagues, responding to emails and using various electronic means to complete tasks. Many engineering managers are familiar with electronic platforms, such as wikis and intranets, or even project management software, on which employees share knowledge and updates on the progress of various tasks. Virtual meetings are also common, especially for geographically dispersed workforces whereas employees often participate in virtual/ distributed Communities of Practice (CoPs) through communication media such as discussion groups or chat rooms as support mechanisms (Lesser & Storck, 2001). Similarly, the online learning environment and community should simulate a similar environment. Online learning is ought to enable students to use the electronic platform to respond to emails, address tasks, work in groups, offer peer feedback to formative tasks, socialize and more.

The principle remains the same with a daily workplace: a group whose members regularly engage in sharing and learning towards accomplishing set targets. The idea of a students' working community somehow simulates the so-called Communities of Practice (CoPs), which are popular in contemporary workplaces (e.g. Rennstam & Kärreman, 2020). Simulating a COP offers opportunities for learning in a diversified environment as online cohorts are often comprised by international students, offering a mix of experience rooted in different cultures, ideas, national backgrounds, religions, genders, abilities, disabilities, sexual orientations, socioeconomic backgrounds, personality types and learning styles – all of them co-existing within an intellectually capacious environment. The essence here is that postgraduate online learning should not be different from daily life at work - while preparing students for the industry and career advancement.

Likewise, social constructivism abounds with community features and social interconnection where knowledge is constructed through social negotiation (Aviv et al., 2003; De Laat et al., 2006). Learning occurs while individuals interact 'with more knowledgeable members of a community within specific social, cultural, and historical contexts' (Kong & Pearson, 2002: 2, cited in Boling et al, 2012: 119). Overall, high levels of interaction result to a positive attitude and greater satisfaction among learners in online education (Desai et al., 2009; Mooney, 2008). This process is enhanced further by the instructor, as explained in the next section. The high level of engagement mitigates the lower richness and emotional engagement of the online environment.

Instructor as Facilitator

Based on social constructivism, education is instructor-led so that learners are guided through the content and learning activities (Rhode, 2009). The instructor, nevertheless, is a facilitator rather than a surrogate teacher or conveyor of information (Smolin & Lawless, 2003). Students are asked to work on several activities, such as formative teamwork, while being observed by the instructor. This transition from conveyor of information to facilitator mirrors the evolution of leadership from authoritarian, to transactional and now servant leadership. The zeitgeist of the time cuts across education and the workplace, allowing the learners to be treated in a way they are familiar with. At the same time, instructional design should be designed to cultivate a sense of community among the learners (Desai et al, 2009). Fostering a sense of community in online courses helps minimize feelings of isolation (Rovai, 2002a), improve student performance and lead to deeper (Hulon, 2013) as well as better learning (Epp et al, 2017).

Furthermore, the social online communities and CoPs, which have been discussed in the previous section, are of great importance as they bring together the intellectual capital of individual members (Sultanovaet al, 2018; Samadi, 2016). If we apply this observation to an online learning community, the analogy refers to the intellectual capital of students. When registering for a post-graduate course, each engineering student or practitioner arrives with previous knowledge, which has been accumulated through experiential learning and technical expertise. The instructor's challenge (and duty) here is to identify the students' intellectual capital and extract it for the benefit of the learners' community. In other words, the instructional design and instructor's facilitation should cultivate an open environment of knowledge transfer, where students can exchange ideas, share previous experiences and collaborate to address certain targets (as we explain in the model later).

Also, the instructor, as a facilitator, should keep an eye on the discussion forums in a systematic manner and make them part of measurement, as we explain in the proposed model later. Unlike the objectivist model, where instructors answer students' questions directly, the constructivism model supports social interaction in discussion forums (Piccoli et al, 2001). A classroom discussion signifies that students' role is more active, compared to passively taking in information. There are various types of classroom discussions, varying from purpose, content and format. Student discussions and the creation of meaning through talking have always been helpful within the confines of brick-and-mortar classrooms (Larson & Keiper, 2016). With the progress of technology, discussion forums became an integral part of online classes too, both synchronous and asynchronous. In fact, online discussion forums signify the evolution of correspondence education to online learning through the diffusion of computers and the internet.

Several accounts exist, exploring in detail the effectiveness of group discussions as a tool for participatory citizenship, knowledge sharing and collaborative learning (Gokhale, 1995; McKeachie, 1996; Jeong, 2003). However, it is also worth mentioning that some learners voice their annoyance with discussion forums. Discussion forums are not necessarily everyone's preferred way of learning (Chiu & Hew, 2018). Therefore, instructors and instructional designers should make sure that discussion forums are meaningful and authentic. As we suggest in our measurement model later, discussion forums should have a clear purpose and be able to provide students with experiences at various levels of cognitive processing (Chiu & Mok, 2017) and behavioural actions that challenge their higher-order cognitive skills (Gillespie, 1998) Several aspects of learning take place in the discussion forums, including the so called 'Learning by Doing', which is discussed below.

Learning by Doing

Formative tasks should be in place to leverage hands-on learning. This is because students learn in more breadth and depth when they participate and do things (Barton, 1995). Towards this end, a series of structured formative tasks should be offered to students on a weekly basis, asking them to do things. A discussion forum, for example, where students work in groups to answer case studies or solve problems, signify that students' role is more active, compared to passively taking in information. It is a form of participatory citizenship, knowledge sharing and collaborative learning (Gokhale, 1995; Jeong, 2003).

With formative tasks, students are busy formulating their own policies, inventing their own product solutions, starting up companies from scratch and adopting different roles. Also, students may be asked to complete reflective tasks, where they are expected to explore and integrate theories, models and tools in their own cultural contexts and workplaces. These are occasions where learners reflect on, and construct their own meaning and knowledge from their own life and workplace experiences (Jonassen, 2004). Reflective tasks enable students to test established theories and practices to new environments. Then, students are expected to upload and discuss their findings with the other members of the cohort on the learning platform. Uploading their work on the platform is important as motivation increases when students realize that their work will be displayed (Bonk & Dennen, 1999). This is the point where knowledge is generated, shared with and retained among the members of the online community. Then, as explained below, students may also offer feedback to each other. Peer feedback seems to be an essential component of effective teaching and learning (Boling et al, 2011).

Systematic Feedback as Part of Continuous Assessment

As mentioned earlier, while students accumulate knowledge through working on formative tasks and interacting with their colleagues, it is important to offer and receive feedback systematically, by both peers and instructors. The benefits of peer feedback have long been explored and theorized by various researchers in the past (Gielen et al, 2010; Hanrahan & Isaacs, 2001; Boud, 2014, 1989, 1995). Commentators argue that sometimes peer judgements are even more important than instructor appraisals (Boud et al, 2014).

Instructors, of course, should be cautious when directing students to offer feedback to each other. Cheng et al (2015) explain that negative peer judgments may evoke negative emotional responses and affect participation. Similarly, Shute (2008) communicates that negative peer feedback may disrupt student performance. In other words, the type, content and style of peer feedback is decisive in students' overall leaning experience (Cho & MacArthur, 2010; Gielen et al, 2010; Nelson & Schunn, 2009). Towards this end, Cheng et al. (2015) make a distinction between the various feedback types to conclude that cognitive feedback (e.g. direct correction) is often more helpful than affective feedback (e.g. praising comments) and metacognitive feedback (e.g. reflecting comments). Therefore, students (in the same way as instructors), should be aware of, or receive training (ideally during the online induction) on the proper way of offering as well as accepting peer feedback.

Moreover, 'systematic instructor feedback', is an equally essential element in the process of learning. Instructor feedback should be offered to all students' contributions, both formative and summative. Offering feedback at the end of the course simply to inform students where they have gone wrong is unlikely to enhance student quality learning (Grosjean & Sork, 2007). Feedback should be continuous; a motivating two-way dialogue; in the form of constructive critique rather than criticism. Often, formative evaluations are the only opportunity for students to receive instructor feedback and improve their weaknesses prior to submitting a summative assessment. Ongoing feedback helps keep students updated on their progress and serves as a foundation for formal assessments. While the benefits of continuous feedback are clear it should not be excessive as this may demotivate the learner and make them feel that they have limited initiative.

ANALYSIS: A MEASUREMENT MODEL FOR ASSISTING STRATEGIC LEARNING

A conceptual model is a set of concepts with propositions, used to describe an event or process (Meredith, 1992). As explained by Delbridge and Fiss (2013), this typically involves creating a nomological network around the focal concept (in our case: social constructivism learning techniques) and delineating it through different relationships and key variables (in our case: metrics). By measuring his or her performance, a student or instructor will be able to evaluate whether certain goals are met. To carry out this process, a learning provider needs to put in place performance metrics, together with the constructivism learning aspects discussed in the literature review, namely: 1. interaction, 2. learning by doing, 3. peer work and 4. systematic feedback.

Figure 2. Performance metrics and constructivist learning

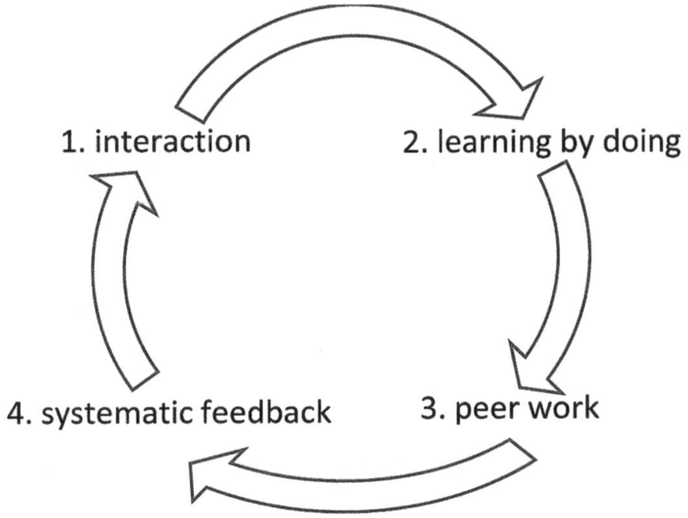

To support the proactive performance measurement (PM) process, learning providers and learners need to decide which aspects of their learning performance needs to be measured. To do that, they need to have in place three basic components: Objectives, Critical Success Factors (CSFs) and Key Performance indicators (KPIs). These three components form a pyramid, with each of the components supporting the outcome of the one above. For each objective, we can have a three CSFs, and then a number of KPIs, for each CSF.

Figure 3. Objectives, CSFs, KPIs - three components forming a pyramid.
Source: Harvard Business Press.

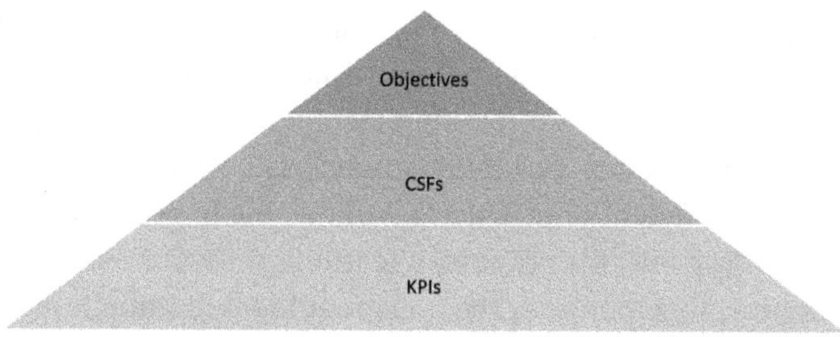

Let us assume at this point that a course is about to commence in the upcoming semester. The course leader needs to have in advance all three components in place. Below, we offer an example for each component, for the course's first week:

Level 1: Objectives

An objective directly affects the learning strategy and if achieved, would help address important tasks and milestones. One objective, for instance, could be the following: *understanding what each weekly effort is required by the students to complete their group formative tasks (techniques: learning by doing; interaction; peer feedback) including guided reading.* By having a clear understanding of what is expected, students will complete the course's first milestone, which is to complete the weekly guided reading and interact with their group members to develop knowledge and submit their first teamwork by the end of the week. However, the objective is too broad. Therefore, to help students formulate a clear understanding, the course leader needs to develop Critical Success Factors, as those appearing below.

Level 2: Critical Success Factors (CSFs)

For each objective, course leaders identify a number of CSFs, which are the actions a student must take to achieve each objective. These could take the following form:

CSF1: *students need to spend time reading the core material for week 1;*
CSF2: *students need to invest time in peer interaction to prepare and submit the formative task;*
CSF3: *students need to improve punctuality (e.g. to inspect their message inbox and respond to peers on a timed manner) and communication skills (to exchange messages and drafting peer work).*

Level 3: Key Performance Indicators (KPIs) for Each CSF

KPIs help course leaders measure how well their students do on different tasks, such as the peer interaction and formative tasks we mentioned earlier. Module leaders need a set of KPIs, that supports the CSFs and Objective we discussed earlier. But where will this performance data come from? The report logs on each learning platform (e.g. a Moodle platform) contain very rich data. Platform logs accumulate information based on several metrics, which reflect and quantify the effort of students, as follows:

KPI1: *percentage of time spent by each student in peer interaction on the platform;*
KPI2: *percentage of time spent on platform core readings;*
KPI3: *percentage of time spent on the group forum, exchanging ideas, suggesting, offering feedback and editing the group's script (to also unearth students' intellectual capital as suggested earlier, we need meaningful and authentic discussion forums);*
KPI4: *to what extent/percentage do we value adequate levels of platform attendance*

KPIs indicate the kind of data you will need to gather. This data is easy to export by the platform, its reliable and offers an immediate understanding of the big picture.

TARGETS

After deciding what aspects of learning performance require measurement; defining objectives; identifying critical success factors (CSFs) for each objective; and defining key performance indicators (KPIs) for each CSF; it's time to set targets for each KPI. Targets represent the performance students want to see on each KPI. Targets are what the students can see and may consider as the benchmark of their performance. It's important, nevertheless, to highlight that the whole process of measurement and setting targets work when:

- There is an understanding of the type of target being used.
- The targets fit the context and the instructional design of the university and programme (in line with the techniques presented in the literature review).
- It is important to pay attention to the way that the targets are framed as this has an impact on the student behavior - especially when it comes to formative tasks and student interaction, which are often optional and voluntary as they do not count towards the final grade.
- Clarity and commitment are provided (a target that is not understood by students is not likely to be very effective) (Source: adapted by Harvard Business Press, 2017)

The development of targets by instructors can be informed by previously gathered data (e.g. platform percentages by previous cohorts and semesters). There are several types of targets. For example, a student-friendly approach would be to set a three-point range for certain targets: minimum, moderate, and "stretch":

- *A minimum target of 3 hours of peer interaction per week*
- *A moderate target of 5 hours of peer interaction per week (possibly for students who want to become the deputy leader of the group).*
- *A stretch target of 8 hours of peer interaction per week (possibly for students who want to assume the leadership of the group).*

This kind of 'target range' is friendly as it offers students an option. It also feels more manageable (especially the minimum targets). At the same time, the range enables those students with ambitions and genuine commitment to reach moderate levels, even stretch targets, and assume leadership responsibilities in the team.

Another type of targets, is the Benchmarking. For example, one target could be: to complete the peer task successfully, you need to use information from the following sources:

1. One Trade Association Report;
2. The Core textbook;
3. One Government Report;
4. Two academic articles.

A common benchmarking target for online students is to require three posts per week, each utilizing two sources. Although these are just a few examples, it's important to note that targets should be realistic. Their purpose is to inspire students and facilitate their learning (Harvard Business Press).

DISCUSSION

The measurement tools presented above, together with the constructivism aspects discussed in the literature review, are just a few examples that can be used each week as part of the particular measurement model. In reality, there are 1000s of measures to choose from (Harvard Business Press). For each of the components, different examples can be used to develop an intellectually capacious student experience. Also, the process can be repeated each week, with different components. All components together, comprise a model that can serve as a proactive navigation instrument – so that instructors and learners know what is expected and what is required. In addition, the model can serve as a warning alarm and prevent several problems (Reichheld, 1996). By tracking performance, both students and instructors can identify and address problems such as declining student attendance, low student performance, attrition and more.

Important in this process, prior to developing each objective, is to decide as a course leader what you hope to accomplish by measuring performance and how you may better serve the needs of your students. In fact, targets can be discussed and set with students as part of a feedback process whereas Higher Education institutions should be in constant consultation with students. It is important to promote student feedback surveys through cultivating a feedback culture and mindset. Equally important is to make sure that survey findings are taken into serious consideration; and inform student training in regard to required skills and knowledge – which in turn may become part of the measurement proccss. Students should have the opportunity to give teachers feedback on their learning experience and performance measurement, so instructors know how learning strategies helped them achieve intended outcomes; and adapt instructional design where necessary. Student feedback is even more critical in online education because of the challenges posed by the distance concerning those involved (Miliam et al. 2004).

The benefits of a measurement model are numerous. The measurement model in combination with an instructor acting as a facilitator creates a shared vision in the classroom. Looking to measure aspects of learning help instructors and students understand the dimensions of each aspect; give them a sense of control; and enhances students' self-development (Harvard Business Press, 2009). Online learning through social constructivism is a complex activity. Applying metrics to this process reduces the complexity of each step in the process. In addition, such a model lessens some of the special difficulties and unique conditions that online learners and providers are likely to encounter. Firstly, as stated earlier, distance may result to feelings of isolation (Rovai, 2002a). Secondly, synchronous connections are not always successful as online students may live in areas of the world with poor connectivity. Another downside is that students located in different time zones may be at work or may have to wake up in the middle of the night to attend. Hence, in the same way that poor connectivity, different time zones and other technical difficulties emerge in our daily life at home and work, they can also happen in an international online studying environment. Moreover, students may also be disrupted or confused by abstract, subjective or negative peer feedback (Shute, 2008). The model we suggest, nevertheless, aims at lessening those difficulties by offering objective benchmarks that can be set in both synchronous and asynchronous environments. Other typical challenges are reduced trust, reduced satisfaction and lower engagement. Encouraging and guiding a structured engagement improves satisfaction (Cheng et al. 2016). Adapting the learning objectives and forms of delivery to the topic also improves satisfaction (Cheng et al. 2019). Lastly a high level of structured peer interaction in online learning builds trust (Cheng et al. 2017).

These benchmarks collect valuable data. While this chapter covers the direct benefits of this data to the current module being delivered the data can also be used by Artificial Intelligence and data technologies to inform the further development of the modules.

CONCLUSION

This chapter applies the principles of measurement on the elements of social constructivism for effective online learning. Performance measurement as part of strategizing is not new in complex systems such as universities. Universities' emphasis on effectiveness and efficiency has naturally led on to a thorough examination of measures of performance (e.g. Ghannadpour et al, 2018). Likewise, the pedagogy discussed in the foregoing analysis is not novel (its origins are dated back to the pre-World War II era). However, the measurement model we propose and some of its learning strategies are new. For instance, so far Intellectual Capital has been linked to education through studies focusing on university managers and faculty (e.g. Mariani et al, 2018; Sultanovaet al, 2018; Samadi, 2016; Secundo et al, 2018). This chapter, nevertheless, is to our knowledge the first attempt to apply the notion of intellectual capital on learners. Unearthing students' intellectual capital, enables the development of a knowledge-competence base, which in turn contributes to the total value of learning experience. A major requirement for the instructor is to be able to identify and improve the visibility of intangible and non-material qualities of students (such as previous professional experience), to capture and unpack this knowledge, and utilize it through teamwork, social networking and learning by doing. This learning process can be enhanced further by the proposed measurement model.

Such a model, adds rationale and reduces complexity as it measures certain aspects of learning. At the same time, it gives students a sense of control as it provides them with steppingstones towards addressing important learning outcomes and milestones. This, in turn, enhances students' self-development.

Moreover, the model also aims to create a shared vision in the module and smoothen some of the difficulties that often emerge in the online learning process, such as studying through distance, dealing with different time zones, dealing with unclear or subjective expectations, limited body language, reduced trust and limited engagement.

In closing, we support that there is no best model. No model is final and no model is static. On the contrary, each pedagogy model is subject to the ever-changing forces in the environment, such as changes in technology, infrastructure, learning resources and always in the light of students' and employers' changing needs. The misconceptions and myths related to the difficulty of teaching and learning online gradually disappear and are replaced by learners' realities – not with a one size fit all product but through quality pedagogical improvements. Relying on functional relationships between social constructivism and measurement, the proposed model contributes to literature by proposing a quantified methodology for learning.

REFERENCES

Alraimi, K. M., Zo, H., & Ciganek, A. P. (2015). Understanding the MOOCs continuance: The role of openness and reputation. *Computers & Education*, *80*, 28–38. doi:10.1016/j.compedu.2014.08.006

Anastasiades, P. (2002a). In Towards the global information society: the enactment of a regulatory framework as a factor of transparency and social cohesion. Lecture Notes in Computer Science, 2510, 527–535. doi:10.1007/3-540-36087-5_62

Aviv, R., Erlich, Z., Ravid, G., & Geva, A. (2003). Network analysis of knowledge construction in asynchronous learning networks. *Journal of Asynchronous Learning Networks*, *7*(3), 1–23.

Barton, J. (1995). Conducting Effective Classroom Discussions. *Journal of Reading*, *38*(5), 346–350.

Batiz-Lazo, B., & Efthymiou, L. (2016). *The Book of Payments: Historical and Contemporary Views on the Cashless Economy*. Palgrave McMillan. doi:10.1057/978-1-137-60231-2

Boling, E., Hough, M., Krinsky, H., Saleem, H., & Stevens, M. (2012). Cutting the distance in distance education: Perspectives on what promotes positive, online learning experiences. *The Internet and Higher Education*, *15*(2), 118–126. doi:10.1016/j.iheduc.2011.11.006

Bonk, C. J., & Dennen, V. P. (1999). Teaching on Web: With a little help from my pedagogical friends. *Journal of Computing in Higher Education*, *11*(1), 3–28. doi:10.1007/BF02940840

Boud, D. (1989). The Role of Self-Assessment in Student Grading. *Assessment and Evaluation in Higher Education*, (14), 20–30.

Boud, D. (1995). *Enhancing learning through self-assessment* (1st ed.). Kogan Page.

Boud, D., Cohen, R., & Sampson, J. (2014). *Peer Learning in higher Education: learning from and with each other*. Routledge. doi:10.4324/9781315042565

Cheng, K. H., Liang, J. C., & Tsai, C. C. (2015). Examining the role of feedback messages in undergraduate students' writing performance during an online peer assessment activity. *The Internet and Higher Education*, *25*(4), 78–84. doi:10.1016/j.iheduc.2015.02.001

Cheng, X., Fu, S., Han, Y., & Zarifis, A. (2017). Investigating the Individual Trust and School Performance in Semi-Virtual Collaboration Groups. *Information Technology & People*, *30*(3), 691–707. doi:10.1108/ITP-01-2016-0024

Cheng, X., Su, L., & Zarifis, A. (2019). Designing a talents training model for cross-border e-commerce: A mixed approach of problem-based learning with social media. *Electronic Commerce Research*, *19*(4), 801–822. doi:10.100710660-019-09341-y

Cheng, X., Wang, X., Huang, J., & Zarifis, A. (2016). An Experimental Study of Satisfaction Response: Evaluation of Online Collaborative Learning. *International Review of Research in Open and Distributed Learning*, *17*(1), 60–78. doi:10.19173/irrodl.v17i1.2110

Chiu, T. K. F., & Hew, K. F. (2018). Factors influencing peer learning and performance in MOOC asynchronous online discussion forum. *Australasian Journal of Educational Technology*, *34*(4). Advance online publication. doi:10.14742/ajet.3240

Chiu, T. K. F., & Mok, I. A. C. (2017). Learner expertise and mathematics different order thinking skills in multimedia learning. *Computers & Education*, *107*, 147–164. doi:10.1016/j.compedu.2017.01.008

Cho, K., & MacArthur, C. (2010). Student revision with peer and expert reviewing. *Learning and Instruction*, *20*(4), 328–338. doi:10.1016/j.learninstruc.2009.08.006

Conrad, D. (2006). E-Learning and social change: An apparent contradiction. In M. Beaudoin (Ed.), Perspectives on higher education in the digital age, (pp. 21–33). New York: Nova Science Publishers.

Cornelissen, J. (2017). Editor's comments: Developing propositions, a process model, or a typology? Addressing the challenges of writing theory without a boilerplate. *Academy of Management Review*, *42*(1), 1–9. doi:10.5465/amr.2016.0196

De Laat, M., Lally, V., Lipponen, L., & Simons, R.-J. (2006). Analysing student engagement with learning and tutoring activities in networked learning communities: A multi-method approach. *International Journal of Web Based Communities*, *2*(4), 394–412. doi:10.1504/IJWBC.2006.011766

Delbridge, R., & Fiss, P. C. (2013). Editors' comments: Styles of theorizing and the social organization of knowledge. *Academy of Management Review*, *38*(3), 325–331. doi:10.5465/amr.2013.0085

Desai, M., Hart, J., & Richards, T. (2009). E-learning: Paradigm shift in education. *Education*, *129*(2), 327–334.

Dewey, J. (1938). *Experience and education*. Collier Macmillan Publishers.

Efthymiou, L. (2018). Worker body-art in upper-market hotels: Neither accepted, nor prohibited. *International Journal of Hospitality Management*, *74*, 99–108. doi:10.1016/j.ijhm.2018.02.012

Efthymiou, L., Orphanidou, Y., & Panayiotou, G. (2019). The Latest from the Tourism Front: Technology, Innovation and Disruption. *The European Financial Review*, *4*(5), 39-43. Online at: https://www.europeanfinancialreview.com/the-latest-from-the-tourism-front-technology-innovation-and-disruption/

Epp, C. D., Phirangee, K., & Hewitt, J. (2017). Student Actions and Community in Online Courses: The Roles Played By Course Length and Facilitation Method. *Online Learning*, *21*(4), 53–77.

Ghannadpour, S., Jokar, M., & Makui, A. (2018). Fuzzy analytical network process logic for performance measurement system of e-learning centers of universities. *Journal of Industrial and Systems Engineering*, *11*(3), 261–280.

Gielen, S., Peeters, E., Dochy, F., Onghena, P., & Struyven, K. (2010). Improving the effectiveness of peer feedback for learning. *Learning and Instruction*, *20*(4), 304–315. doi:10.1016/j.learninstruc.2009.08.007

Gillespie, F. (1998). Instructional design for the new technologies. In K. H. Gillespie (Ed.), *The Impact of Technology on Faculty Development, Life, and Work. New Directions for Teaching and Learning, 76*. Jossey-Bass. doi:10.1002/tl.7603

Gilson, L. L., & Goldberg, C. B. (2015). Editors' comment: So, what is a conceptual paper? *Group & Organization Management*, *40*(2), 127–130. doi:10.1177/1059601115576425

Gokhale, A. A. (Fall 1995). *Collaborative Learning Enhances Critical Thinking Journal of Technology Education, 7(1)* (M. Sanders, Ed.).

Grosjean, G., & Sork, T. J. (2007). Going online: Uploading learning to the virtual classroom. *New Directions for Adult and Continuing Education*, *113*(113), 13–24. doi:10.1002/ace.243

Guilar, J., & Loring, A. (2008). Dialogue and community in online learning: Lessons from Royal Roads University. *Journal of Distance Education*, *22*(3), 19–40.

Hanrahan, S. J., & Isaacs, G. (2001). Assessing self- and peer-assessment: The students' views. *Higher Education Research & Development*, *20*(1), 53–69. doi:10.1080/07294360123776

Harvard Business Press. (2009). Measuring Performance: Expert Solutions to Everyday Challenges. *Harvard Business School Press Pocket Mentor*. Online at: https://store.hbr.org/product/measuring-performance/13280

Huang, M. H., & Rust, R. T. (2018). Artificial intelligence in service. *Journal of Service Research*, *21*(2), 155–172. doi:10.1177/1094670517752459

Hulon, S. (2013). Face-to-face discussions versus online threaded discussions: Can we have the best of both worlds? In R. McBride & M. Searson (Eds.), *Proceedings of Society for Information Technology & Teacher Education International Conference 2013*, (pp. 577-582). Chesapeake, VA: Association for the Advancement of Computing in Education.

IATA. (2018). *Partner-taught and Self-study*. Online at: https://www.iata.org/training/delivery/Pages/distance-learning.aspx

Institute, A. P. (2016). *Implementing a Performance Measurement Framework at Essex Police*. Online at: https://www.bernardmarr.com/default.asp?contentID=1060

Jaakkola, E. (2020). *Designing conceptual articles: four approaches*. AMS Rev.

Jeong, A. (2003). The Sequential Analysis of Group Interaction and Critical Thinking in Online Threaded Discussions. *American Journal of Distance Education*, *17*(1), 25–43. doi:10.1207/S15389286AJDE1701_3

Jonassen, D. H. (Ed.). (2004). *Handbook of Research for Educational Communications and Technology* (2nd ed.). Lawrence Erlbaum Associates.

Keegan, D. (1996). *Foundations of distance education* (3rd ed.). Routledge.

Kong & Pearson. (2002). 2, cited in Boling, E., Hough, M., Krinsky, H., Saleem, H. & Stevens, M. (2012). Cutting the distance in distance education: Perspectives on what promotes positive, online learning experiences. *The Internet and Higher Education*, *15*(2), 118–126.

Larson, B. E., & Keiper, T. A. (2016). *Classroom Discussion and Threaded Electronic Discussion: Learning In Two Arenas*. Online at: https://www.citejournal.org/volume-2/issue-1-02/general/classroom-discussion-and-threaded-electronic-discussion-learning-in-two-arenas/

Leontyev, A., & Baranov, D. (2013). Massive open online courses in chemistry: A comparative overview of platforms and features. *Journal of Chemical Education*, *90*(11), 1533–1539. doi:10.1021/ed400283x

Lesser, E., & Storck, J. (2001). Communities of Practice and Organizational Performance. February 2001. *IBM Systems Journal*, *40*(4), 831–841. doi:10.1147j.404.0831

Lin, J., Cantoni, L., & Murphy, J. (2018). MOOCs in tourism and hospitality: A review. *Journal of Teaching in Travel & Tourism*, *18*(3), 217–235. doi:10.1080/15313220.2018.1428505

Mariani, G., Carlesi, A. & Scarfò, A. (2018). Academic spinoffs as a value driver for intellectual capital. The case of the University of Pisa. *Journal of Intellectual Capital, 19*(1), 202-226.

McIsaac, M. S., & Gunawardena, C. N. I. (1996). Distance education. In D. H. Jonassen (Ed.), Handbook of Research for Educational Communications and Technology, (pp. 403-437). New York: Simon & Shuster Macmillan.

Meredith, J. (1992). Theory Building through Conceptual Methods. *International Journal of Operations & Production Management*, *13*(5), 3–11. doi:10.1108/01443579310028120

Miliam, J., Voorhees, R. A., & Bedard-Voorhees, A. (2004). Assessment of online education: Policies, practices and recommendations. *New Directions for Community Colleges, 126*(126), 73–85. doi:10.1002/cc.156

Mooney, C. (2008). IT on the campuses: What the future holds. *The Chronicle of Higher Education, 54*(30), 89.

Moore, M. G., & Kearsley, G. (2011). *Distance Education: A systems view of online learning*. Cengage Learning.

Nelson, M. M., & Schunn, C. D. (2009). The nature of feedback: How different types of peer feedback affect writing performance. *Instructional Science, 37*(4), 375–401. doi:10.100711251-008-9053-x

Payne, A., Frow, P., & Eggert, A. (2017). The customer value proposition: Evolution, development, and application in marketing. *Journal of the Academy of Marketing Science, 45*(4), 467–489. doi:10.100711747-017-0523-z

Piccoli, G., Ahmad, R., & Ives, B. (2001). Web-based virtual learning environments: A research framework and a preliminary assessment of effectiveness in basic IT skills training. *Management Information Systems Quarterly, 25*(4), 401–426. doi:10.2307/3250989

Ragin, C. C. (1994). *Constructing social research*. Pine Forge Press.

Reichheld, F. (1996). Learning from Customer Defections. *Harvard Business Review, 29*(6), 755–756.

Rennstam, J., & Kärreman, D. (2019). Understanding control in communities of practice: Constructive disobedience in a high-tech firm. *Human Relations, 73*(6), 864–890. doi:10.1177/0018726719843588

Rhode, J. (2009). Interaction Equivalency in Self-Paced Online Learning Environments: An Exploration of Learner Preferences. *The International Review of Research in Open and Distributed Learning, 10*(1). Advance online publication. doi:10.19173/irrodl.v10i1.603

Rovai, A. P. (2002a). Building sense of community at a distance. *International Review of Research in Open and Distance Learning, 3*(1). Online at: http://www.irrodl.org/index.php/irrodl/article/view/79/152

Samadi, A. (2016). The Impact of Intellectual Capital on Students' Creativity from Teachers'. *International Journal of Humanities and Cultural Studies, 2*(4), 1584–1591.

Secundo, G., Massaro, M., Bagnoli, C., & Dumay, J. (2018). Intellectual capital management in the fourth stage of IC research: A critical case study in University settings. *Journal of Intellectual Capital*, *19*(1), 157–177. doi:10.1108/JIC-11-2016-0113

Shute, V. J. (2008). Focus on formative feedback. *Review of Educational Research*, *78*(1), 153–189. doi:10.3102/0034654307313795

Smolin, L. I., & Lawless, K. A. (2003). Becoming literate in the technological age: New responsibilities and tools for teachers. *The Reading Teacher*, *56*(6), 570–577.

Spector, J. M., Merrill, M. D., Merrienboer, J. V., & Driscoll, M. P. (2008). *Handbook of research on educational communications and technology* (3rd ed.). Lawrence Erlbaum Associates.

Sultanova, G., Svyatov, S., & Ussenbayev, N. (2018). Transmitting competencies at universities in Kazakhstan: Intellectual capital of teachers. *Journal of Intellectual Capital*, *19*(1), 112–134. doi:10.1108/JIC-04-2017-0058

Sveiby, K. E. (2010). *Methods for measuring intangible assets*. Available at: www.sveiby.com/portals/0/articles/IntangibleMethods.htm

Tham, C. M., & Werner, J. M. (2005). Designing and evaluating e-learning in higher education: A review and recommendations. *Journal of Leadership & Organizational Studies*, *11*(2), 15–25. doi:10.1177/107179190501100203

Vrasidas, C. (2000). Constructivism Versus Objectivism: Implications for Interaction, Course Deign, and Evaluation in Distance Education. *International Journal of Educational Telecommunications*, *6*(4), 339–362.

Vygotsky, L. S. (1978). *Mind in society*. Harvard University Press.

Wagner, E. D. (1994). In support of a functional definition of interaction. *American Journal of Distance Education*, *8*(2), 6–29. doi:10.1080/08923649409526852

Whiteley, T. R. (2006). Using the Socratic Method and Bloom's Taxonomy of the Cognitive Domain to Enhance Online Discussion, Critical Thinking and Student Learning. *Developments in Business Simulation and Experiential Learning*, *33*, 65–70.

Chapter 2
Teaching a Technical Writing and Research Course to Engineering Students:
Recommendations for Curriculum Reform

Anna-Maria Andreou
Ministry of Education, Culture, Sports, and Youth, Cyprus

EXECUTIVE SUMMARY

Training in communication skills is considered extremely important in the engineering profession. However, educational organisations and most specifically engineering programs and departments have often been criticised for failing to adequately prepare engineering students for the situations they will face in the workplace. This chapter describes a technical writing and research course that is offered as a required course to engineering students and analyses the advantages and limitations of the course pointing to changes in the course development that will enable students to perform successfully as communicators in the workplace.

INTRODUCTION

In today's changing technology landscape and information, new modes of communication and global interconnectedness, engineering education has a very important role to play in equipping students with professional skills and competencies which are considered imperative in the workforce. Engineering graduates should

DOI: 10.4018/978-1-7998-4063-3.ch002

possess not only deep technical and broad professional skills and be problem solvers, but also a number of professional skills (i.e., "soft skills") important for the profession such as effective communication and persuasion, teamwork, and understanding the business realities within a global and societal context, ethics and professionalism. According to Johansen and Johansen (2007), among the attributes that technology leaders of the future should have are the ability to collaborate with people in large groups, use different persuasive techniques depending on the social context and media space they work in, design content for public consumption and modification, and filter meaningful information, patterns and commonalities from ever-growing streams of data.

Thus, the workplace demands new engineers to be technically qualified, flexible, and dynamic thinkers because economic progress and achievements are dependent on their science and engineering talent. Professional and non-technical skills (e.g., communication, team, design, ethics, leadership, organization) are undoubtedly highly prized. Nevertheless, engineering students continue to experience problems of speaking and writing for engineering purposes notwithstanding the importance of the afore mentioned non-technical skills and competencies that engineering graduates should possess, or the need to engage more efficiently in instruction related to these skills.

This chapter begins by highlighting the importance that engineering graduates themselves place on oral and written technical communication for successful practice on the workplace and lays out various reasons which account for this gap, including curricular slotting and time constraints which do not allow room for new nontechnical requirements or collaboration between engineering and technical communication faculty, and competency gaps related to speaking and writing.

The sections that follow describe the technical communication and research course as it is integrated in the engineering program at the University of Nicosia and analyse its advantages and limitations for preparing engineering graduates for their future communication activity. Then some recommendations for changes are offered that could contribute in improving technical communication for engineers in light of workplace and alumni survey results which show that both students and the industry, one of the primary customers of the university, ask to hire graduates with deep technical and broad professional skills.

BACKGROUND

The importance of communication skills for the engineering profession has been acknowledged and affirmed in employer and alumni surveys that aim to identify the role that communication skills play in professional life. For example, results from a

U.S. Industry-Government Roundtable for enhancing Engineering Education survey of 420 engineers and engineer managers revealed that survey participants considered communication one of the top 11 skills for success in the engineering workplace (Lang et al., 1999). These results were confirming the findings of two previous studies (Evans et al, 1993; Benefield et al., 1997), which rated communication skills second and among the top fifth key success factors respectively. Similarly, results of a 2001 survey showed that former engineering students from three different classes ranked technical communication and interpersonal skills highest of all the skills listed for importance in professional life (Tryggvason et al., 2001). Pinelli et al. (1995) surveyed the amount of time 2,355 aerospace engineers and scientists spent communicating technical information at work. Results showed that producing their own technical communication varied from 19.6 to 23.3 hours and responding to technical communication from others from 14.9 to 19.6 hours. Sageev and Romanowski (2001) also found in a survey targeting 1500 newly graduated engineers who attended the School of Engineering and Applied Sciences of the University at Buffalo between 1994 and 1996 that 208 respondents spent 64% of their overall work time on communication such as writing (32%), oral presentations (10%) and other oral discussions such as conveying ideas and concepts in meetings (22%). They also found that the amount of hours used on technical communication increased as they advanced professionally.

Reave (2004) examined instruction in technical communication skills by surveying four main methods of instruction in 73 top-ranked U.S and Canadian engineering schools during the school year 2003-2004: required courses, integrated instruction, elective courses and engineering communication centres. Required courses include technical writing such as project proposals and reports, product instructions, lab reports, and product specifications, and sometimes oral presentations and business communication such as letter and memo writing, CVs and cover letters. Some schools include instruction in team communication to meet the Accreditation Board for Engineering and Technology/Canadian Engineering Accreditation Board (ABET/CAAB) criterion 3.d, 'ability to function on multidisciplinary teams.' Integrated courses are those which provide communication instruction as a substitute or supplement into core engineering courses such a Design and Communication. Elective courses cover technical communication skills in speaking and writing and are often more advanced versions of the introductory level courses. Finally, engineering communication or writing centres offer individualised instruction in writing and speaking. The authors conclude that for engineering communication skills to be developed, communication programs in engineering schools need to offer a good foundation in business and technical communication through a stand-alone course or integrated instruction followed by further instruction throughout the curriculum accomplished through team teaching or partnering of communication and engineering faculty collaborating on instructional design, delivery and evaluation. However, few engineering schools have developed programs that distribute instruction in writing and communication throughout the student's course work and 13 schools do very little to help students meet the workplace needs of the profession.

A large gap between practice needs and engineering graduates' communication skills is also reported by Seat, Parsons and Poppen (2001) in an industry study by the Society of Manufacturing Engineers reporting 14 competency gaps such as working with a team, interacting with diverse populations, and writing and speaking. In a 2003 study of Canadian engineering firms by Heuchan and Archibald, engineers in 30 mining companies rated verbal communication, written communication, technical writing and presentations among the five highest in importance subjects out of 29 different curriculum subjects. However, the rated quality of educational preparation they had received in these subjects showed a gap indicating their dissatisfaction with the level of preparation in these subjects. Similar results were shown in alumni surveys (McGourty, 1999; Riley, Furth and Zellmer, 2000; Core Curriculum, 2003; Sageev and Romanowski, 2001) in which the consensus was that oral and written communication skills rated very high in importance, but the level of preparation they had received in this area was inadequate.

The results were similar in another survey of about 260 young professional electrical engineers (VDE 2003), who felt well prepared as regards their technical training, but reported competency gaps in their presentation skills, marketing know-how and management and negotiating skills.

According to Becker (2006), Siemens managers are evaluated regularly based on their performance in skills shown in table 1 (Becker p.266).

Table 1. Additional skills expected of engineers in an international company

Work-related Skills	Management-related Skills
Management of knowledge/IT Flexibility in adapting to change Focus on results and client needs Focus on international opportunities Dedication to life-long learning Systematic and logical approach Ability to communicate Commercial/ðnancial knowledge/expertise Application of relevant knowledge/expertise Curiosity, creativity, cross-border thinking Good command of business English Ability to work in teams	Providing leadership and vision Commitment to ethical and social responsibilities Commitment to principles of sustainability Leadership/motivation of people Management of projects/events Team building skills; multidisciplinary/cultural Management of self/time Convincing communication, verbal and written Learning, developing and improving Interest for trends in society Act and behave according to circumstances

Source: (Becker, 2006)

He reports that 'employers have long since pleaded for university reform that would address the actual deðciencies and is in line with international trends' (p.267).

The communication competences that the engineer of the future should have pose a challenge for engineering schools of today to design the curriculum and pedagogical approaches that will prepare engineering graduates for technical work: communicate effectively, use creative problem-solving strategies and critical thinking skills, and so on. An influential voice in this regard is the Boeing Company, and its Desired Attributes of an Engineer. In addition to a good understanding of engineering science fundamentals, these attributes include: design and manufacturing processes, system perspective, understanding the context of engineering (including economics, history, environment, customer and societal needs), ethical standards, teamwork, critical and creative thinking, curiosity and a desire to learn for life (The Boeing Company).

In response to complaints from the industries that hire engineering program graduates, the Accreditation Board for Engineering and Technology (ABET), which is responsible for the quality of American engineering and technology programs, adopted the Engineering Criteria 2000 (EC 2000), by which engineering programs must document student outcomes, that is what has been learned, in a variety of skills areas. One requirement is that communication skills should be integrated with technical work throughout the curriculum, not only in communication courses alone, so that students can have multiple opportunities to develop all their skills in a way that best reflects the professional situations where technical skills and communication skills are used in concert. Since the implementation of EC 2000, there has been a shift to academic and practicing engineers alliances to determine curricula, which in turn have held engineering programs accountable for educating students to work in teams, communicate effectively, and 'understand the economic, social, environmental and international context of their professional activities'.

In Europe, quality assurance (QA) procedures were introduced after the creation of the European Higher Education Area (EHEA), commonly known as the Bologna process, to respond to the rapidly changing economy due to globalization in the 1990s which in turn has led international companies to have different requirements regarding academic qualifications. For example, since the advancement of a 'consecutive' university education, based on two cycles, undergraduate and graduate degrees in Germany, a great number of higher education German programmes have been transformed to abide by the consecutive model and have been already accredited, especially in engineering. In the UK, France and Portugal, discipline-specific accreditation has preceded the introduction of the overall QA procedures. In some countries such as Italy programme accreditation procedures are being developed in parallel with the Bologna process. In Cyprus, ETEK (Technical Chamber of Cyprus) is the official Body that offers professional accreditation to engineers in Cyprus. All the engineering department's graduates are registered by ETEK as Professional

Mechanical Engineers. Additionally, trans-national recognition of academic and professional qualifications to strengthen the position of the European engineer in the global employment market has been addressed by the EUR-ACE Framework Standard that provides a common reference framework as the basis for the award of a common European quality label involving the active participation of national accreditation agencies and leading to a multilateral mutual recognition agreement awarded by six Accreditation Agencies in six different countries (namely, Engineering Council-UK, Engineers Ireland; Order of Engineers, Portugal; RAEE, Russia; CTI, France; ASIIN, Germany)(Augusti, 2009).

Notwithstanding the generally quite high quality of European engineering programmes within the context of a global standard and the efforts made towards innovating and reforming engineering curricula, the evolving consensus is that the students' deficiency in a number of professional skills (i.e., "soft skills") important for the profession such as effective communication and conceptual and intuitive understanding of the world where engineering is practiced, communication, team, design, ethics, leadership, and organization, weakens the position of the European engineer in the global employment market and prompt action.

MAIN FOCUS OF THE CHAPTER

Science and engineering play an important role in promoting economic growth and prosperity, improving the standard of living and the quality of life, boosting productivity and reducing unemployment. Engineering plays an important role in other sectors of life including healthcare and medicine and is constantly changing.

With the aim to better prepare engineering graduates to adapt to new challenges and for every contingency that they may encounter in their careers, students are taught basic engineering principles and concepts, on specialized topics ranging from power and automation systems, communications and fibre optics to computer networks, microwave, and Very-large-scale integration (VLSI) system design. The need to be independent thinkers is facilitated by requiring them to produce independent types of work, perform high-tech projects, write technical reports, perform research on cutting-edge technology, and present their work before classmates and faculty members.

The department of Engineering at the University of Nicosia offers undergraduate studies (BSc) in five engineering disciplines: Civil and Environmental Engineering, Computer Engineering, Electrical Engineering, Mechanical Engineering, and Oil and Gas Engineering. During the four-year programmes, graduates develop a deep conceptual and intuitive understanding of basic engineering principles and concepts as well as laboratory experience on specialised topics ranging from earthquake resistant structures and state of the art drilling and exploration of hydrocarbons to autonomous cyber physical systems, unmanned low-emission/electric vehicles and high-speed wireless communication networks. The curriculum leading to the degree of Bachelor of Science (BSc) in any of the five engineering programs aims to qualify students for employment in a variety of industry sectors ranging from academic and research organisations to government agencies and private companies, or to pursue further advanced post-graduate studies at the level of Masters and/or Doctorate degree. The major engineering courses of the Bachelor's program contain, apart from theory and fundamental principles, elements of analysis and design, as well as significant laboratory work. The duration of the Bachelor of Science (BSc) programs, for a full-time student, is 4 years, and it requires a minimum of 240 ECTS credits for graduation. Students in the two engineering programs take courses of basic Science (e.g., Physics), Mathematics, Computer Science, Information Technology (IT), Electronics, Design, Business, English Language, and Liberal Arts in order to prepare for a professional career in industry or to pursue advanced post-graduate studies in engineering or a related field. The process of system analysis and design is emphasized throughout all years of study and, as a matter of fact, projects are included in all the major theoretical and laboratory courses. This provides the theoretical foundation and the practical skills for our students to successfully complete a demanding final year project at the end of their undergraduate journey.

Furthermore, the department offers post-graduate studies (MSc and PhD) in Electrical Engineering and in Oil, Gas and Energy Engineering. The duration of the Master of Science (MSc) programs, for a full-time student, is 1,5 years, and it requires a minimum of 90 ECTS credits for graduation whereas for the Doctor of Philosophy (PhD) program well-qualified applicants can complete their program in 3 to 8 years depending on the mode of study (full time or part time) and after they have completed 180 ECTS credits.

All programs conform to the academic guidelines set by international accreditation bodies such as the Accreditation Board for Engineering and Technology (ABET) in the United States. The curricula have been reviewed by the Cyprus Scientific and Technical Chamber (ETEK) and graduates can register as licensed Engineers with ETEK.

The goals are that analytical and theoretical knowledge balanced with the iterative process of designing, predicting performance, building and testing, should be taught from the earliest stages of the curriculum; therefore, the pathway to achieve those goals was formulated to include major and elective courses of theoretical and applied nature. For example, students in the undergraduate engineering programs take courses of basic Science (e.g., Physics), Mathematics, Computer Science, Information Technology (IT), Electronics, Design, Business, English Language, and Liberal Arts in order to prepare for a professional career in industry or to pursue advanced post-graduate studies in engineering or a related field.

A proposed breakdown of required and elective courses in the Bachelor of Science in Computer Engineering per semester includes a language requirement, ENGL-101 English Composition, during the first semester of year 1, BADM-332 Technical Writing and Research in the first semester of year 4, and a Liberal Arts elective which may include COMM-200 business and Professional Communications out of a of five other liberal arts electives in the second semester of year 4. Students can plan when to choose each course after they have studied carefully their academic path and consulted with their academic advisor and the Department of Academic Affairs. In addition, they may have to take additional English courses depending on their proficiency in the English language.

The main objectives of English Composition are to help students produce academic papers (research papers, thesis etc.), develop their critical research skills, distinguishing between sources, synthesizing information, as well as understand referencing and citation rules. Students also work on academic written material and develop academic vocabulary and processes and strategies to produce academic papers.

The main objectives of BADM-332 Technical Writing and Research course are to provide students who are preparing for a career in business, social and applied sciences, engineering, and other technically and empirically oriented professions with knowledge for applying written communication skills at the workplace, make students aware of the principles and mechanics of formal and informal technical writing, thoroughly discuss the various types of technical writing (e.g. resume, business letters with focus on job application letters, memos and reports), discuss the methods of primary research data collection (qualitative and quantitative), the structure of primary research papers and various documentation styles. After completion of the course students are expected to be able to:

1. create resumes, job application letters and memos
2. distinguish the usefulness and formats of different types of technical writing
3. write clearly, concisely, ethically and in a non-discriminatory way for maximum communication effectiveness
4. solve problems of biased language and wordiness in the form of short exercises
5. distinguish various writing, organizational and editing techniques
6. analyse and critically evaluate written materials composed by others
7. identify the processes of conducting primary research and the writing of primary research papers, including
    a. various quantitative and qualitative methods of data collection
    b. the structure of a primary research paper
    c. various styles of source citation and documentation

   d. ethics in data collection and writing
8. demonstrate increased self-awareness and self-confidence in writing through collaborative interaction with peers and through independent study

The main objectives of COMM-200 Business and Professional Communication course are to: provide students with the basic principles of effective communication, equip students with the ability to develop their presentation skills, empower students to develop confidence when speaking before an audience through improving their verbal and nonverbal language skills, develop students' listening skills required for effective communication, help students differentiate between giving informative and persuasive speeches, and engage students in group discussions to improve their interpersonal and intercultural communication skills, which are also key aspects in achieving effective communication.

In the next section some problems that result from the implementation of technical communication in the engineering curricula are described followed by recommendations for improvement.

ISSUES, CONTROVERSIES, PROBLEMS

Problem 1: Technical communication instruction is too late

Clearly, the engineering program involves one or more language and expression course(s)early on in the program, however instruction in technical communication is expected to be taken in the fall semester of the fourth year and then the elective course in the spring semester of the fourth year. Students can therefore practise their writing and communication skills throughout the engineering programme, however these courses do not provide any instruction in technical communication. Even though the course can be complemented by a Liberal Arts elective taken in the following semester and that they both involve extensive written and oral communication practice, instruction that is offered in the fourth year of the programme fails to reap the benefits of repeated practice and feedback through distribution of instruction throughout the program. Taking a single technical communication course earlier on in the program would not necessarily improve a student's performance or constitute the program successful in educating engineers; rather, students must be provided with continuous opportunities to develop their technical communication skills through a more integrated approach to engineering and technical communication. The idea is not new and various approaches have been well documented and implemented in engineering schools (Reave, 2004) including that of communication across the curriculum (CAC)(Reave, 2004). According to Reave, in a CAC program instruction is distributed throughout the student's course work by either creating portfolios of writing as well as project reports, lab reports, technical memos, letters of application, poster presentations and other assignments from a variety of courses, or through a series of required communication courses for engineers in a 4-year CAC program. The advantage of CAC programs is that they allow time for several important content areas to be covered which are often not included due to lack of time, and 'they keep communication consistently in the foreground" (p. 467), providing settings that will enable students to prepare for workplace communication.

Problem 2: Technical communication instruction is too little

A single semester in technical instruction is not sufficient to prepare the engineering students for workplace writing. Another problem of the BADM-332 Technical Writing and Research course is that it blends instruction of the various types of technical writing with research data collection methodology and structuring of primary research papers in just 13 weeks, i.e. 39 hours. The combined effect is that it helps students with modes of communication they will use in their future careers while at the same time will provide writing and research instruction that is important for the final year project. As a result, instruction in technical communication skills is too little. Instructors have to cover important material in a very limited period of time. To expand communication instruction the BADM-332 Technical Writing and Research course should be two separate stand-alone courses so that technical communication that represents true engineering communication with clients, vendors, managers or investors is given full attention in the instruction. If this substantial investment of time cannot be realised due to curricular time boundaries, expansion of technical communication instruction can be achieved through integrated instruction as a supplement to BADM-332. More specifically, engineering professors can integrate writing into their courses by giving students assignments that reflect work demands, for example writing reports and giving presentations. An integrated approach to engineering and communication would prevent obstacles that discourage engineering faculty from making the effort including lack of complete fluency or appropriate qualifications or scholarship (Reave, 2004) in the field of communication, the cumbersome and time consuming process of reading, grading, providing feedback, and evaluating assignments, the development of effective assignments that emphasize engineering concepts while at the same time encourage good writing, investing time and energy on communication neglecting their own research or other important technical content, and no institutional or technical communication department support to help them accomplish their goals (Williams, 2001). More electives could also be offered such as the existing course in Business Communication, which addresses report writing and intercultural communication, or new elective courses could be created to meet the need for instruction in intercultural technical communication examining engineering and professional communication in different countries.

Problem 3: Collaboration between technical communication instructors and engineers

Technical communication faculty members could collaborate with the engineering professors on areas of the course where writing might prove useful and together develop tasks, plan instructional delivery, support and evaluation. At UNIC partnering could be implemented between communication and engineering faculty on the senior year project (ECE-492), which aims at teaching students important research techniques and practices, introducing students to practical engineering design, utilizing theoretical knowledge and engineering tools/techniques acquired throughout the years in order to design, build, and test their idea in a laboratory environment, promoting team work and practical experience in a multi-disciplinary environment, teaching students how to write proper reports and how to present their work in front of their colleagues, and promoting engineering ethics and respect to the environment and society. The advantage of this partnership is that students can receive specialised instruction through lectures given by both communication and engineering faculty and be supervised by both engineering and communication instructors on final projects. It will also prevent allocation of most of the classroom time and attention on engineering design rather than communication or writing. Of course, these partnerships will entail changes in tenure and reward systems as well as teaching loads, but will ultimately provide a competitive advantage to students, faculty and their universities.

Problem 4: Exemplification

The importance of clarity, conciseness and accuracy in various communication channels are stressed in technical communication coursebooks and are primary objectives of effective technical communication both grammatically and contextually. According to Gerson and Gerson (2012) technical communicators can achieve clarity by supplying specific details in fewer words. Yet, results of study by Ktoridou and Doukanari (2015) indicate a lack of exemplification skills in their posts on a technology-related blog. More specifically, using discourse analysis and informal conversational interviews, the authors set out to examine the nature of students' technical writing skills such as relevance, exemplification, appropriate use of jargon and mode of writing, on the blog UNIC Tech Freaks that was developed as a group project by students enrolled in an undergraduate course, MIS-220 Technologies for the Social Web, during spring 2015 semester at the University of Nicosia, Cyprus. Postings on the blog were about technology innovations in smartphones, tablets, laptops, gadgets and operating systems which attracted not only the student-developers but also other users sharing similar interests to participate in the interactions on, for example, product features, specifications, practicality, aesthetics and cost, thus creating a community of young blog users. Informal conversational interviews from the developers of the blog revealed their satisfaction with the project, stimulating interaction, involvement and learning through knowledge exchange and content with their peers. They also affirmed the social and learning importance of student generated content (SGC) by providing the students with a sense of identity as members of that particular community of users. However, one of the most salient features of technical writing - exemplification - was inadequate as in the authors' words (p. 293) "Most of the comments are vague and lack exemplification -a skill that is considered essential in technical writing. They express an opinion, without providing any details or explanations to justify the comments they make".

While understanding that the reasons for these limitations in the students' postings were not the focus of the study, it is important to acknowledge that the Unic Tech Freaks blog generated informal conversational written expression and jargon on a particular technological device that was described as motivating and educational. According to Freedman and Artemeva (1998), teaching engineering writing from the notion of situated learning, where learning is seen as a social process accomplished through collaboration, the blog postings allow for the kind of interaction among peers that is recognised as a powerful tool for learning. Set within the context of social constructionist genre theory (Bakhtin, 1986; Miller, 1984), which interprets learning to understand and manipulate the genres of written communication in one's field as essential to professional success, the blog postings enrich the discursive context within which students operate. To improve exemplification, technical communication teachers could ask students as part of assessment in the technical communication class to keep posting on a blog they develop their comments on the subject-matter of other engineering courses they are taking concurrently with the technical communication course, or comment on their peers' postings. The instructor will also participate in the discussion answering and asking questions, commenting on students' postings and providing assistance and feedback. The goal is for students to take full advantages of the communal collaborative ambience created by the peers interactions and the instructor's feedback and assistance in the course of the exchange of the written utterances so that students gradually learn how to enrich their interactions with examples, and details appropriate for both the engineering-content courses and to workplace writing.

SOLUTIONS AND RECOMMENDATIONS

Recommendations for further improvement of engineering education at the University of Nicosia may not be viewed as particularly novel but are grounded on tried and true ingredients that make them work at a variety of institutions and which include:

- A communication-across-the -curriculum (CAC) approach to instruction that is distributed throughout the student's course work by asking students to create portfolios of writing as well as project reports, lab reports, technical memos, letters of application, poster presentations and other assignments from a variety of engineering courses, or through a series of required communication courses for engineers in a 4-year CAC program in order to expand communication instruction.
- Integrating writing into engineering courses by giving students assignments that reflect work demands, for example writing reports and giving presentations.
- Partnering between communication and engineering faculty on the senior year project (ECE-492) so that students can be offered situated learning and individualised feedback and evaluation by both engineering and communication instructors on final projects.
- Continued technical communication and research skills development in the craft of teaching should be readily available and highly encouraged through the participation on a student developed technology-related blog where students will comment on the subject-matter of the engineering courses they are taking and the instructor can also participate by answering and asking questions, commenting on students' postings and providing feedback. This in turn will drive continuous and systemic improvements in students' exemplification skills, which are essential in technical writing.

FUTURE RESEARCH DIRECTIONS

In the previous sections we argue that schools should utilize integrated instruction in which communication specialists and engineering professors collaborate. Engineering and technical communication faculty who work together can create new courses where blended technical and communication content introduce students to the modes of communication, for example, lab reports, project proposals, and design reports that are appropriate to the field of engineering. To this end, future research should investigate how collaboration between engineering and technical communication faculty can drive continuous and systemic improvements in the light of classroom experience and new research and theory.

CONCLUSION

The fact is that the world needs top-notch engineers in ever-increasing quantities to address not only local problems but grand engineering challenges, to support economic as well as social development. The engineering graduate of today in theory and in reality should possess very good knowledge of engineering fundamentals and be technically competent and possess technical communication skills so that s/he can cope when exposed to business constraints when designing a solution, or when s/he has to work in multidisciplinary, multi-national, multi-discipline teams. Professional and non-technical skills (e.g., communication, team, design, ethics, leadership, organization) are highly prized, yet the typical engineering curriculum is bursting with required technical courses, thus there can often be a limited interest in offering more ECTS credits for technical writing by the engineering department.

Of course, implementing changes to the engineering curricula is not an easy feat and requires a more extensive survey/study of which steps should be taken to promote an integrated approach to instruction while at the same time balancing research with teaching loads of faculty, promoting faculty training and faculty partnering.

REFERENCES

ABET Engineering Accreditation Commission. (2010). *Criteria for accrediting engineering programs*. ABET.

Augusti, G. (2009). EUR-ACE: the European Accreditation system of engineering education and its global context. In *Engineering Education Quality Assurance* (pp. 41–49). Springer. doi:10.1007/978-1-4419-0555-0_3

Bakhtin, M., & Ghāsemipour, G. (2011). The problem of speech genres. *Literary Criticism*, *4*(15), 114–136.

Becker, F. S. (2006). Globalization, curricula reform and the consequences for engineers working in an international company. *European Journal of Engineering Education*, *31*(3), 261–272. doi:10.1080/03043790600644749

Benefield, L. D., Trentham, L. L., Khodadaki, K., & Walker, W. F. (1997). Quality improvement in a college engineering instructional program. *Journal of Engineering Education*, *86*(1), 57–68. doi:10.1002/j.2168-9830.1997.tb00266.x

Core Curriculum and Undergraduate Services Committee. (2003, June). *Proposal for revising the general education component of engineering undergraduate curricula.* Retrieved November 12, 2003, from www.eng.ohio-state.edu/faculty/forms/Eng_GEC_Proposal.pdf

Evans, D. L., Beakley, G. C., Crouch, P. E., & Yamaguchi, G. T. (1993). Attributes of engineering graduates and their impact on curriculum design. *Journal of Engineering Education, 82*(4), 203–211. doi:10.1002/j.2168-9830.1993.tb01075.x

Freedman, A., & Artemeva, N. (1998). Learning to teach writing to engineers. *Canadian Journal for Studies in Discourse and Writing/Rédactologie, 14*(1).

Gerson, S. J., & Gerson, S. M. (2012). *Technical communication: Process and product*. Prentice Hall.

Heuchan, J., & Archibald, J. (2003). *Industry and higher education: Meeting the needs of the mining engineering sector*. Retrieved November 27, 2003, from http://mine.queensu.ca/people/faculty/Archibald/Surv'03Paper.pdf

Johansen, R., & Johansen, B. (2007). *Get there early: Sensing the future to compete in the present.* Berrett-Koehler Publishers.

Ktoridou, D., & Doukanari, E. (2015, November). Student generated content in higher education technology-related blogs. In *2015 international conference on interactive Mobile communication technologies and learning (IMCL)* (pp. 289-295). IEEE.

Lang, J. D., Cruse, S., McVey, F. D., & McMasters, J. (1999). Industry expectations of new engineers: A survey to assist curriculum designers. *Journal of Engineering Education, 88*(1), 43–51. doi:10.1002/j.2168-9830.1999.tb00410.x

McGourty, J. (1999). *Alumni say communication skills are essential.* Retrieved March 3, 2003, from www.engineering.columba.edu/news/archive/engnews_s99/survey.html

Miller, C. R. (1984). Genre as social action. *The Quarterly Journal of Speech, 70*(2), 151–167. doi:10.1080/00335638409383686

Pinelli, T. E., Barclay, R. O., Keene, M. L., Kennedy, J. M., & Hecht, L. F. (1995). From student to entry-level professional: Examining the role of language and written communications in the reacculturation of aerospace engineering students. *Technical Communication (Washington), 42*, 492–507.

Reave, L. (2004). Technical communication instruction in engineering schools: A survey of top-ranked US and Canadian programs. *Journal of Business and Technical Communication, 18*(4), 452–490. doi:10.1177/1050651904267068

Riley, L. A., Furth, P. M., & Zellmer, J. T. (2000). *Assessing our engineering alumni: Determinants of success in the workplace.* Retrieved November 12, 2003, from www.ece.nmsu.edu/pfurth/papers/00/ASEEABET.pdf

Sageev, P., & Romanowski, C. J. (2001). A message from recent engineering graduates in the marketplace: Results of a survey on technical communication skills. *Journal of Engineering Education, 90*(4), 685–697. doi:10.1002/j.2168-9830.2001.tb00660.x

Seat, E., Parsons, J. R., & Poppen, W. A. (2001). Enabling engineering performance skills: A program to teach communication, leadership, and teamwork. *Journal of Engineering Education, 90*(1), 1–12. doi:10.1002/j.2168-9830.2001.tb00561.x

The Boeing Company. (n.d.). http://www.boeing.com

Tryggvason, G., Thouless, M., Dutta, D., Ceccio, S. L., & Tilbury, D. M. (2001). The new mechanical engineering curriculum at the University of Michigan. *Journal of Engineering Education, 90*(3), 437–448. doi:10.1002/j.2168-9830.2001.tb00624.x

VDE. (2003). *Young Professionals.* Available online at: www.vde.com/vde_en/

Williams, J. M. (2001). Transformations in Technical Communication Pedagogy: Engineering, Writing, and the ABET Engineering Criteria 2000. *Technical Communication Quarterly, 10*(2), 149–167. doi:10.120715427625tcq1002_3

ADDITIONAL READING

Kalogerou, V., & Andreou, A. M. (2015). The role of quality assurance in detecting lacunae in the activities of the Ecostar project and implementing suitable modifications. *INTED2015 Proceedings*, 7864-7869.

Kalogerou, V., Andreou, A. M., & Maoz, H. (2015). The development and implementation of the 'Advanced Thinking Skills' course with the support of an Extended Communication Platform. In *Strategies for Teaching in the XXI Century* (pp. 201–206). Cambridge Scholars Publishing. https://books.google.com.cy/books?isbn=1443884855

KEY TERMS AND DEFINITIONS

Bologna Process: A series of agreements made between various European countries set out in June 1999 to ensure common standards, comparability, and reliability of higher education qualifications across Europe.

CAC Program: A Communication Across the Curriculum Program aims to incorporate oral and written assignments and activities into courses across disciplines.

Discourse Analysis: A research method for analysing written and spoken language use in relation to its social context.

ECTS: The European Credit Transfer and Accumulation System aims to make national education systems more comparable internationally.

ETEK: The Cyprus Scientific and Technical Chamber (ETEK) is the statutory Technical Advisor to the State and is the umbrella organisation for all Cypriot Engineers.

Soft Skills: Soft skills are non-technical skills and include communication, problem-solving, creativity, adaptability, and work ethic.

Technical Communication: A field that helps users to convey scientific, engineering, or other technical information.

Chapter 3
The Design and Redesign of an Online Socio-Constructivist Course on Engineering Management:
The Role of Learning Scenarios and Learning Analytics

Mary Grammatikou
National Technical University of Athens, Greece

Nadia Sansone
https://orcid.org/0000-0002-3413-7983
Sapienza University of Rome, Italy

Dimitris Pantazatos
https://orcid.org/0000-0002-0924-1260
National Technical University of Athens, Greece

Donatella Cesareni
Sapienza University of Rome, Italy

Vasilis Maglaris
National Technical University of Athens, Greece

EXECUTIVE SUMMARY

This chapter focuses on the design and redesign of an online engineering management course, based on a socio-constructivist approach. At first, the theoretical and contextual premises will be presented with a focus on the suggested teaching and

DOI: 10.4018/978-1-7998-4063-3.ch003

learning methods to acquire domain-related knowledge and crucial skills and on the importance of learning scenario to support an effective learning design. After the background introduction, a user case will be described, focusing on the course online environment and its tools, on the proposed pedagogical strategies and above all, on how instructors can obtain and analyze useful educational data from various sources. Finally, some redesign recommendations will be provided to better use educational data for continuous course improvement.

INTRODUCTION

Engineering management is a specialized field of management concerned with the engineering sector. Reflecting industry demand for management-focused engineers, a growing number of specialized engineering management degrees are available to help develop the knowledge and skills needed for these roles. During an engineering management course, students develop industrial engineering skills, knowledge and expertise, alongside knowledge of business and management techniques, strategies and concerns. Plus, one of the main goals of higher education, whatever the domain, is to ensure that students acquire useful skills to achieve success not only in their studies but also in their future career and life in general. Through specific techniques and educational strategies, students should learn to act, study and work intentionally and effectively, individually or together with others, in authentic contexts, solving complex problems and creating new solutions and new knowledge. Based on these outcomes and core competencies, instructors should design their courses, focusing not just on the courseware, rather on the whole teaching/learning experience. Once a course is fully designed, according to the specific goals and theoretical approaches the instructor bear in mind, it comes into practice and the design process starts again.

The scope of this chapter is to examine the design and redesign of an Engineering Management course, by following some research questions:

- Which are the core skills that a student and future worker should develop to be successful in his/her career?
- Which teaching and learning strategies could better support the development of the identified skills?
- Which methods a teacher should follow to design and redesign an effective course?

To answer these questions, at first, the theoretical and contextual premises will be presented, with a particular focus on the suggested teaching and learning methods suggested to acquire nowadays crucial skills. After the background introduction, we will describe our use case, with clear attention to the online environments and its tools, on the pedagogical strategies and, above all, on how instructors can obtain and analyze useful educational data from various sources, according to the Learning Analytics framework. Finally, some redesign recommendations will be provided to better use educational data for continuous course improvement.

THEORETICAL BACKGROUND

One of the main goals of higher education, whatever the chosen knowledge-domain, is to ensure that students acquire useful skills to achieve success not only in their studies but also in their future career and life in general (Sansone, Cesareni, Ligorio, Bortolotti & Buglass, 2019). In every age, the "useful skills" are defined according to the context in which those skills should be mobilized (Le Boterf, 1994). The context in which today's students live and work is that of a high technological impacted knowledge work society. In other words, it is a society where knowledge and technology represent two "inextricably linked" factors in any educational and professional context (Scardamalia, Bransford, Kozma, and Quellmalz, 2012, p. 234) and where uncertainty and rapidity of change are intrinsic dimensions. These features encourage the development of specific skills and consequently imply the need for educational agencies to review curricula and pedagogical practices, as also highlighted by the international community (OECD, 2013). A comprehensive list of 21st-century skills has been provided by Binkley and colleagues (2012) who identified ten skills grouped into four clusters: ways of thinking (e.g., creativity and innovation, critical thinking, problem solving, decision making, learning to learn and metacognition); ways of working (e.g., communication and collaboration – teamwork); tools for working (e.g., information literacy, ICT literacy), and living in the world (e.g., citizenship; life and career; personal & social responsibility). Equipped with these skills, tomorrow citizens should learn to act, study and work intentionally and effectively, individually or together with others, in authentic contexts, solving complex problems and creating new solutions and new knowledge. As mentioned, this means also becoming more aware of the cultural power technologies play which should not be considered merely as a content delivery mechanism, but rather as artifacts able to support human capacity for developing new culture (Engeström & Escalante, 1996; Wartfosky, 1973), therefore academic and educational contexts are called upon to develop positive, cultural models of how to use this technology.

It is a long time now that literature points out the many advantages of online learning (Clark & Mayer, 2007; Horton, 2006). Yet, the path that led from the first Distance Education era to the different models of nowadays e-learning universities has seen, over the years, a change of perspective and theoretical paradigm. The first online universities were based on a behaviorist paradigm, centered on a model of a teacher who defines, elaborates and prepares the contents, and of a student as the one who merely acquires and memorizes them. The socio-constructivist paradigm overturns this vision and instead proposes a model of a teacher who guides the course with appropriate stimuli, and of a student as one who builds his/her own knowledge by interacting with objects, meanings, and people. In today's online universities these two ways of understanding knowledge and learning can be reproduced. Inspired by a behaviorist paradigm, attention will be placed on a good organization and presentation of knowledge, taking care of the quality of the lessons and the organization of the contents, so that they are extremely comprehensible and easily assimilable by the students. These are certainly important aspects, but if the authors believe as Vygotsky (1978) that knowledge is created in interaction, the authors need to go further and, inspired by the socio-constructivist model, propose activities that lead to building complex knowledge, through interaction with teachers, tutors, and peers. The lack of face-to-face interaction seems still to question the validity and effectiveness of online education: learners' involvement tends to decrease, because of rarefaction of meta-communication aspects (Waztlawick, Beavin, & Jackson, 1971), of mutual feedback between learners and teachers, and more generally of participants' social presence (Rovai, 2002). These limits are even more evident in online courses with little or no interaction between the participants.

The socio-constructivist approach combines technology and educational contexts to promote collaborative, constructive, and meaningful learning through students' active role (Jonassen, 2006; Scardamalia & Bereiter, 2006). To reach this goal it is, however, necessary to re-think the educational contexts and to plan activities based on specific teaching and learning strategies, such as project-based learning, flipped classroom and so on. Particularly, the learning path should be structured around activities strongly based on a) specific pedagogical goals, b) students' active involvement, and c) procedures reified through the use of learning artifacts (Dillenbourg, 2002). In this way, the acquisition and participation metaphors of learning (Sfard, 1998) are embedded in the knowledge creation metaphor, going beyond two traditional dichotomies: individual versus social processes, and conceptual knowledge versus social practices, both needed to foster collaborative creativity (Paavola, Lipponen, and Hakkarainen 2004). And here it stands the mediation of tools that, in distance teaching, obviously have a central role. Already in 2003, Trentin defined e-learning as "a way of using information and communication technology to support teaching/learning processes based on electronic content delivery, on learning active and/or collaborative"(Trentin, 2003, p. 47). Technologies, rather than replace the teacher and/or the classroom, enrich the learning environment and allow experiences otherwise impossible, since they mediate and stimulate new ways of thinking, thus boosting our capabilities (Sansone, Bortolotti, Buglass, 2016).

DESIGNING AND REDESIGNING AN ONLINE COURSE: THE ROLE OF LEARNING SCENARIOS AND LEARNING ANALYTICS

Introducing technologies to support students to take an active role as knowledge builders is a delicate process. Facilitating effective active learning can be complex because academic achievement should involve not only knowledge acquisition, but also meaningful and lasting learning in which learners construct new knowledge, actively participate in learning episodes, and experiment with new skills (Biggs, 2003; Delors, 1997). For technology to support this kind of active learning, teachers must be skilled Learning Designers, able to consider the needs of a range of students and adapt flexibly to technological affordances and innovations, contextual constraints, and resources (Sansone, Cesareni, Bortolotti, Buglass, 2019). Indeed, Mor, Ferguson, and Wasson (2015) suggest that a teachers' intimate understanding of such factors provides them with a means of designing learning opportunities that are "fit for purpose" (p.222). However, teachers should not assume that this knowledge alone will produce effective learning design immediately. Any teacher has his/her own methods, tips, and tricks to note down and remember the lesson plans and the sequence of activities to be carried out in class. These skills are developed during the teacher's career or learnt from senior colleagues. While a large part of our knowledge comes from formally planned learning scenarios, people learn a lot from an informal exchange with fellow learners, with professors, experts, i.e. from exchange within tightly or loosely defined communities (Lave & Wenger, 1991). Nevertheless, some type of formalization is needed to facilitate the design process and make scenarios easy to share and modify (Ligorio, Andriessen, Baker, Knoller, & Tateo, 2009). Based on the chosen outcomes and the targeted core competencies, instructors should design their courses, focusing not just on the courseware, rather on the whole teaching/learning experience. As a result, the sequence of learning activities that learners undertake to attain some learning objectives should be defined, including the resources and support mechanisms required to help learners to complete these activities and their temporary relations (Miao, Sloep, Hummel, Koper, 2009). The whole designing process ends up with a more or less formal description of a pedagogical scenario that may or may not follow an instructional design model. The scenario describes learning objectives, who does what and when, using tools and resources, and outcomes. A pedagogical or learning scenario, as defined by Peter and Vantroys (2005), is a planned description of learning that outlines:

- The sequencing of the activities,
- The learning objects,
- The tools that could be provided by different actors (researchers, teachers, students, parents, experts, etc.).

Finally, a Learning Scenario promotes the use of structured activity-based teaching, which enables the development of skills to be broken down into manageable chunks for the learners (Schneider, Synteta, Frété, Girardin, & Morand, 2003). In this sense, the scenario supports the teacher who wants to innovate their teaching.

Once a course is fully designed, according to the specific goals and theoretical approaches the instructor bear in mind, it comes into practice and the design process starts again. As the ADDIE (Branch, 2009) model suggests, learning design is a circular and never-ending process in which the final stage (naming "Evaluation") could originate a new design, based on the retrieved strengths and weakness. This stage is often performed by following the procedures of the Design-Based Research (DBR, 2003) approach. DBR focuses on the analysis of educational practices by interrogating the planning of innovative interventions and subsequently observing the (not)functioning elements. In this way, limits are identified, and further interventions are re-designed to overcome or address these limits. The aim of DBR is to directly impact practices, while simultaneously stimulating the theoretical progress: the value of a theory is evaluated based on how much its principles improve the practice.

Course design and redesign involve promoting useful and measurable learning outcomes, choosing effective teaching strategies and learning experiences, aligning assessment methods with course learning outcomes, and revising the course syllabus. Specifically, the redesign should be based on various feedback which is obtained, among others, through students' evaluations and from the course outcomes based on the Learning Analytics (LA). LA is the measurement, collection, analysis, and reporting of data about learners and their contexts, for purposes of understanding and optimizing learning and the environments in which it occurs (LAK, 2011). Based on the context in which they are used, LA can be defined as a prediction model, as a generic design framework, as a data-driven decision making, as a process based on educational data or even as an application of Web Analytics. LA can integrate multiple analytics processes that may vary between Learning Management Systems (LMS) or even between the same tools. The majority of the LA data can be visualized either inside an LMS (e.g Moodle, Blackboard) via a dedicated plugin or via the data that can be stored and visualized in an external platform (e.g. Learning Locker). All the collected data can then be analyzed through various procedures like data mining, machine learning, and statistics. Altogether, they compose the Educational Data Mining field seeking to develop and improve methods for exploring the educational data to discover new insights about how people learn in the context of such settings (Educationaldatamining.org, 2020).

The data on which to base the course redesign also depend on the adopted pedagogical approach. As it was mentioned before, in fact, introducing technology in learning can take place in different ways depending on the different pedagogical approaches, moving along a continuum that goes from more transmissive models to more interactionist and constructivist models. From the learning model depends, of course, the type of assessment: to a transmissive pattern generally corresponds a summative assessment, based on the attribution of scores/ratings by the teacher at the end of the path; if, on the other hand, the model envisages learning not only as an acquisition of knowledge, but as an active construction, the assessment method will try to take into account the complex underlying dynamics and observe, rather than the mere results, the processes of construction of knowledge and social participation implemented both at individual and at a group level. From this point of view, the classical assessment systems - oral test, test, written paper - are not enough, as they mainly aim to verify the acquired knowledge and, as such, do not allow to reflect and bring out those processes.

Taken together, the pedagogical approach and its corresponding assessment generate the final amount of data within the digital environments used: from MOOC platforms to Learning Management Systems, from collaborative writing tools to shared drawing boards, from discussion forums to repositories of online resources. Each of these tools hosting activities, functions and roles for individuals and groups to be performed and then observed.

In the following paragraph, a concrete user-case will be presented, based on a socio-constructivist approach and the consequent LA strategies to be considered in order to redesign it.

THE USER CASE: STRUCTURE, CONTENTS, STRATEGIES AND TOOLS

Nowadays, there is a significantly increasing interest in Engineering Management Courses. Engineering Management is a specialized field of management concerned with the engineering sector. Reflecting industry demand for management-focused engineers, a growing number of specialized engineering management degrees are available to help develop the knowledge and skills needed for these roles. The increased interest in Engineering Management Courses is a result of the Fourth Industrial Revolution which is building on the Third, consisted of the digital revolution that has been occurring since the middle of the last century (Xu, David, Kim, 2018). This revolution is characterized by a fusion of technologies that is blurring the lines between the physical, digital, and biological spheres. As a result, there is a need for engineers able to manage more advanced forms of production and knowledge.

During an engineering management course, students develop industrial engineering skills, knowledge and expertise, alongside knowledge of business and management techniques, strategies and concerns. According to the EUR-ACE Framework (ENAEE, 2020), the outcomes of an engineering program should be the following: Knowledge and Understanding, Engineering Analysis, Engineering Design, Investigations, Engineering Practice, and Transferable Skills, which are not subject-specifically defined.

Here it follows the proposed learning scenario for teachers and instructors who want to lead a socio-constructivist online course on Engineering Management aimed at developing the above-mentioned domain-related knowledge and core competencies. The specific scope of this Unit scenario is to introduce post-graduate students to Engineering Project Management. Learners will first acquire theoretical knowledge through webinars, multimedia material (videos), and interactive contents, then they will be requested to discuss and apply that knowledge to collaboratively build a Project Charter based on an engineering project from a provided case-study. The specific learning objectives of this scenario can be summarized as follows:

- Identification of key deliverables based on business requirements while managing customer expectations,
- Development of business and management skills and techniques,
- Experience and development of transversal skills, both cognitive and socio-relational: critical/reflexive competencies, such as reflection on and evaluation of personal and collaborative work, giving and receiving feedback, organizing collaborative work, taking responsibility, evaluating, and managing information.

Starting from the broad socio-constructivist framework inspiring the user case here proposed, our learning scenario is based on various teaching models such as:

- **Project-based Learning (PBL):** In PBL, students work in groups to solve challenging problems that are authentic, curriculum-based, and often interdisciplinary (Gerhana, Mardiyana, Pramudya, 2017). Learners decide how to approach a problem and what activities to pursue. They gather information from a variety of sources and synthesize, analyze, and derive knowledge from it. Their learning is inherently valuable because it is connected to something real and involves adult skills such as collaboration and reflection. In the end, students demonstrate their newly acquired knowledge and are judged by how much they've learned and how well they communicate it. The most prominent skills that can be acquired through this learning model are Critical Thinking & Problem Solving, as well as, Communication & Collaboration.

- **Scenario-based Learning (SBL):** The SBL model uses interactive scenarios to support active learning strategies such as problem-based or case-based learning. It normally involves students working their way through a storyline, usually based around an ill-structured or complex problem, which they are required to solve [ref]. In the process, students must apply their subject knowledge, critical thinking and problem-solving skills in a safe, real-world context. The most prominent skills that can be acquired through this learning model are Critical Thinking & Problem Solving, and Self-Direction.
- **Collaborative Learning (CL):** CL is a learning method based on the interaction within a group of collaborating students which are identified as a Learning Community aiming to achieve a common goal, in terms of creating new knowledge and products. This expression, therefore, refers to a set of principles, techniques and methods of conducting the class based on of which the students deal with the disciplinary study interacting in small groups, in a collaborative, responsible, supportive way and receiving evaluations based on the results obtained as an individual and as a group. Teachers lead activities focusing on students' attention and providing them with the appropriate "know-how". The most important skills that can be acquired through this learning model are Communication & Collaboration, Critical Thinking & Problem Solving, and Information, Media & Technology Literacy.
- **Role Taking (RT):** RT is a specific type of collaboration script (Dillenbourg, 2002), which consists of an explicit didactic contract between a teacher and a group, and is based on the distribution of learning responsibilities during collaborative learning activities. Indeed, collaborative work insists on the commitment of participants that is expressed not only in taking responsibility for their learning but also for that of others (Slavin, 2011; Webb, 1985; Wenger, Trayner & De Laat, 2011). Recently, RT as a didactic practice has seen widespread use in virtual environments (Strijbos, Martens, Jochems, & Broers, 2004; Weinberger, 2008). Many roles can be designed to support constructive interaction and facilitate learning. Following this assumption, each transversal skill could be then promoted, depending on the designed role.
- **Flipped Classroom (FC):** The FC describes a reversal of traditional teaching where students first gain exposure to new material - usually via reading or lecture videos -, and then class time is used to do the harder work of assimilating that knowledge through strategies such as problem-solving,

discussion or debates (Brame, 2019). As a result of the model's emphasis on self-learning, it is possible to achieve the Self-Direction skill along with the Information, Media & Technology Literacy skills, as well as the Critical Thinking & Problem Solving and Communication & Collaboration skills, thanks to their independent work as well as to the group-work.

- **Lecturing Model (LM):** The LM remains crucial, in fact, through benchmark lessons (diSessa & Minstrell, 1998), the teacher offers definitions of concepts, main ideas, and critical issues for each unit. This model is performed online via synchronous ways of communication, namely the webinar (short for Web-based seminar). A webinar is a presentation, lecture, workshop, or seminar that is transmitted over the Web using video conferencing software. A key feature of a Webinar is its interactive elements is the ability to give, receive and discuss information in real-time. Today, many webinar services offer live streaming options or the ability to record the webinars and publish them to other services later. The most important skills that can be acquired through this teaching model are Communication & Critical Thinking, and Information, Media & Technology Literacy.

The course here presented should be hosted by a Learning Management System (LMS), providing both the administration and delivery features of an online course, as well as the tracking and reporting tools which are crucial functionalities to redesign the same course. In an LMS, instructors can create their learning contents providing the main theoretical information by using video recording software or interactive content creation tools, as well as, presentation files, real use cases, external links, etc. Open Educational Resources (OERs) from Open Repositories can also be used as they are increasingly important in the strategy of creating and sharing content (Otero, Pantazatos & Docampo, 2019-A), as well as free content with appropriate copyrights taken from various video portals. The whole set of learning materials can then be saved in a cloud storage service. In the proposed learning scenario, according to the theoretical framework inspiring it, the instructors should also create interactive content (quizzes, presentations, etc.) related to the course topics by using adequate tools. Another type of learning material that can also be used, would be applications based on Augmented and Virtual Reality (AR/VR), which provide an immersive experience. However, students and even teachers may not be so experienced with this kind of application and they will need some time to cope with them (Otero, Pantazatos & Docampo, 2019-B).

Table 1 summarizes some of the environments and tools that can be used for the implementation of this course.

The majority of these can be used either standalone or part of an eLearning Ecosystem. Next Generation Digital Learning Environment (NGDLE) architecture

Table 1. Suggested digital environments and tools

Tool	Type	Description	License
Moodle (Moodle, 2020)	Course Management System	Open-source LMS that can be used for blended learning, distance education, flipped classroom and other e-learning projects in schools, universities, workplaces and other sectors.	GPLv3+
H5P (H5P, 2020)	Content Collaboration Framework	Easy-to-use toolkit, able to provide engaging active learning experiences, which could be used as pre-class or in-class activities.	MIT+
Jitsi Meet (Jitsi Meet, 2000)	Videoconferencing	Open-source WebRTC JavaScript application that uses Jitsi Videobridge to provide high quality, secure and scalable video conferences. It can be easily integrated with a Moodle-like LMS	Apache Licence 2.0
OBS (OBS, 2020)	Software vision mixer, streaming media	Free and open-source software for video recording and live streaming. It can be used offline for video recording	GPLv2
Synchronized eLearning Content (SeLCont) (Introduction to SeLCont, 2020)	Synchronized eLearning Content Toolkit (Pantazatos, Marinos, Giotis, Vekris, Grammatikou & Maglaris, 2018)	It can be used for capturing lectures offline. It provides adjustable HTML5 Web pages of lecture material (ppt's with or without annotations, PDFs, Internet screens, etc.), synchronized with video-audio, stored in a video provider, or on any web server.	GPLv3+

provides a way to integrate various tools and services (Brown, Millichap & Dehoney, 2015). One example of an NGDLE platform is the Up2University Learning Platform (Grammatikou, Pantazatos, Barth, Corleto, Sansone, Zimniewicz, 2020) which is based on the NGDLE, as a way to support a more open, effective and efficient co-design, co-creation and use of digital contents, tools and services specially adapted for personalized, collaborative or experimental learning by students preparing for university. This Learning Platform provides the majority of the tools mentioned above and many more tools, integrated into an innovative ecosystem suitable for students over 12 years old.

After having presented contents, strategies and tools to be adopted in the proposed user-case, here it follows the learning scenario with its activity's workflows (Table 2), that is how actors, contents and services should interact through the digital environment and tools.

FROM LEARNING ANALYTICS TO STUDENTS'

Table 2. The Engineering Management Course Learning Scenario

Phase	Activities	Tools	Timing
Phase 1: Register to the LMS and Introduction to the Course. Introduction to Project Management	• Students register to the course. An Introductory session takes place, covering all the basic aspects of Project Management via a webinar • Recorded lectures related to the project management will also be available for students • Students answer two quizzes related to the webinar topics	• Moodle login form • Jitsi Meet • Moodle Questionnaire • SeLCont	2 weeks (3 hours per week)
Phase 2: Knowledge Acquisition and Group organization	• Students attend a webinar about: o The Organizational Matrix o Basic Human Resources aspects o Project Management Processes • Students answer two quizzes related to the webinar topics • Instructors form random groups composed of 5-6 members. In each group, three roles are assigned: the Project Manager, the Product Responsible, and the Researcher. • A user case based on a real case is distributed to each group	• Jitsi Meet • Moodle Questionnaire • Moodle forum	2 weeks (3 hours per week)
Phase 3: Knowledge Acquisition and Projects Communication	• Students attend a webinar about: o Stakeholders Identification o Managing stakeholders' engagement o Creating sustainable communication channels with the stakeholders • Students are individually requested to accomplish an assignment related to Projects Communication. The assignment is based on the models presented in the webinar (e.g. Salience Model for Stakeholders). Students can use a Moodle dedicated forum to discuss this topic and share ideas and opinions. • Group-works are asked to implement stakeholders' analysis and communication channels for their specific case study. For example, project stakeholders based on their power and interest can be grouped into four categories to design customized communication. Moodle forum and Jitsi Meet can be used for the collaboration between the group members or between the groups.	• Jitsi Meet • Moodle forum • Moodle folder	2 weeks (3 hours per week)
Phase 4: Knowledge Acquisition and Collaborative Project Management	• Students attend a webinar about: o Developing a Project Charter o Project Management Plan o Manage and Control Work • Group-works are asked to structure a complete Project Management Plan for their case study	• Jitsi Meet • Moodle forum • Moodle database	3 weeks (3 hours per week)
Assessment	In each group, the Project Manager presents their product via a short video. Groups are then asked to comment on the other groups' products during a peer-review feedback session. Teachers will present how the real user cases were handled in comparison with the students' actions. Finally, students are asked to answer a survey to express their feedback and appreciation about the course, suggesting possible improvement and highlighting strengths.	• H5P • Moodle Workshop • Moodle feedback	2 weeks (2 hours per week)

ASSESSMENT AND (RE)DESIGN RECOMMENDATIONS

The assessment of students' performance and behavior in an online course is a challenging and emerging task. The vast majority of the LMSs provide LA that can be used for this purpose. According to Lockyer and colleagues (2013), LA offers a method of gathering information on how learners are interacting with learning resources, each other, and their teachers. They also mention that unlike traditional surveys or focus groups, which rely on participants both opting to provide feedback and accurately remembering and reporting past events, LA can capture data on specific, observable behavior in real-time. Although this overcomes data accuracy difficulties, the challenge posed by LA is interpreting the resulting data against pedagogical intents and the local context to evaluate the success or otherwise of a learning activity. The main principle of learning analytics lies in the extraction of useful and actionable information from the large amount of data generated in online learning systems–i.e. LMS log systems – to inform the different learning actors–institutions, instructors, and students –to improve learning processes. While the objective of the application of learning analytics techniques may vary from case to case – prediction of academic success, implementation of early-warning systems, reduction of attrition rates, etc. –, the analysis mostly relies on one source: the data stored on LMS logs (Hernández-García, Acquila-Natale, Chaparro-Peláez, 2017). The vast majority of the LMSs provide a variety of built-in reports based on log data, but they are primarily descriptive in nature -- they tell participants what happened, but not why, and they don't predict outcomes or advise participants on how to improve outcomes. Log entries, while very detailed, are not in themselves descriptive of the learning process. They tell us about the "who," "what," and "when," but not about the "why" or "how well." Much more context is needed around each micro-action to understand and promote a pattern of engagement.

The final implementation of the Learning Analytics should be based on various models according to the organization's needs. Campbel, DeBlois, and Oblinger (2007) proposed a Learning Analytics model that can be characterized as a starting point for implementing Learning Analytics solutions. This model can be described by using a set of questions, grouped in the following five steps:

- **The Capture step**: This step is the one where Machine Learning is involved to gather useful data. In fact, in this step, one has to answer some questions concerning what data is being collected, how frequently is the data collected, where is the data going to be stored, which format is going to be used to represent all events and if the observations securely stored;
- **The Report step**: This step addresses the information flow collected in the first step with the aim of producing some reports. So, the questions concern

who and how frequently will receive the reports, what kind of information needs to be reported and how the reports will be accessed;

- **The Predict step**: This step concerns the prediction algorithms: their accuracy, their prediction, their input data and finally what predictions are reported to stakeholders.
- **The Act step**: This step concerns the actions to be deployed into the learning environment to improve the overall learning process;
- **The Refine step**: In this step, the questions concern reasoning about the results obtained in the previous steps, included their usefulness.

Some other models which are also known for describing how Learning Analytics can be implemented are The Learning Analytics Cycle LA Mode, The 4-dimensions LA Model, and The Teacher-Inquiry Model. However, all these models are strongly data-driven and require systems that have the ability both to collect data, for example with machine learning algorithms and to enable the system stakeholders to provide actions for the improvement of learning processes. Only the Teacher-Inquiry model accomplishes the LA model with the teacher's point of view, integrating all the previous models with a crucial stakeholder (Sciarrone, 2018). The Teacher-Inquiry model tries to combine Teaching Analytics (methods and tools for analyzing and improving the educational design of a lesson before it begins and evaluating the effectiveness of the course delivery) and Learning Analytics (Sergis and Sampson, 2017).

To identify the specific LA on which to anchor the Teacher-Inquiry model, the available learning environment is to be considered. Many Moodle third-party plugins provide descriptive analytics, as well as in other LMSs. Again, these plugins primarily provide descriptive analytics that rely on human judgment to interpret reports and generate predictions and prescriptions. There are features of LA that support this and are being used through the Moodle environment (Machine Learning based model, Static model and Students at risk of dropping out, Upcoming activities due and No teaching).

Regarding the ML-based models, Moodle provides a framework that is based on Supervised Machine Learning. This framework manages the prediction model's life cycle and, for any given prediction model, the framework supports two different modes (Olivé et. al., 2018). The first one automatically extracts training and validation data from the finished courses according to the features and target, defined for the prediction model. It then trains the prediction model and validates its prediction accuracy. This is referred to as the testing mode. The second one uses all the data from the finished courses as training data and the prediction is done for on-going courses. This is referred to as the production mode.

An important standard that should also be taken into consideration to provide descriptive analytics is the Experience API (or xAPI) which is a relatively new specification for learning technology that makes it possible to collect data about a wide range of experiences a person has (online and offline) (xAPI, 2020), since it lifts many of the older SCORM related restrictions. As a result, many types of activities like Mobile learning, simulations, virtual worlds, serious games, real-world activities, experiential learning, social learning, offline learning, and collaborative learning can be recognized and communicated well with the Experience API. The majority of the modern LMSs can support the xAPI standard, including Moodle which provides almost 500 xAPI statements.

In the Table 3 some ways on how an instructor can assess students' performances, competences, and behavior in an LMS system like Moodle are shown.

Table 3. Moodle Learning Analytics

Plugin	Plugin Type	Useful for	Description
Logs	Report	Teachers, Admins, Decision-makers	Filterable log of events
Activity	Report	Teachers	View count of activities in a course
Activity completion, see also Course completion	Report	Teachers	Completion matrix of students and activities
Feedback, Survey, Questionnaire, Workshop	Activity	Teachers, Researchers	Configurable tool
(Quiz) Statistics	Report	Teachers	Student quiz performance report
(Course) Participation	Report	Teachers	Single student's participation in a course
Course overview	Report	Admins, Decision-makers	Comparison of participation across multiple courses
Course completion status	Block	Students	View of student's own completion of activities
Event monitor	Report	Teachers, Admins, Decision-makers	Proactive, focused monitor tool
Engagement Analytics	Block, Report, Activity module	Teachers	Configurable engagement measurement report and block
Course Dedication	Block	Students, Teachers	Estimated time online for students
Grade distribution	report	Teachers	Visualizes the grades of students in a course
Logstore xAPI	Logstore	Admins, Decision-makers, Researchers	xAPI event stream output

(Moodle Learning Analytics, 2020)

As mentioned before, a crucial aspect of a socio-constructivist course is the assessment stage which cannot be mere of a summative type, instead, it requires a continuous analysis of students' participation and activities during the course so to provide them with formative feedback and to collect comprehensive data to support their evaluation. Ongoing monitoring, moreover, helps reflection and guides the students towards better participation in the subsequent activities.

The question that should be asked by every educational expert is how learning analytics can help the teacher to perform such an assessment, including quantitative and qualitative data and analysis. How the learning analytics can support socio-constructivist teaching and learning approaches? Searching for the answers, a final assessment model is proposed for the course here described.

The model considers different aspects of students' learning and participation, by using the different available Moodle LA (Table 4).

Table 4. The Engineering Management Course LA

Activity	LA Plugin	Retrieved Data
Webinar fruition	Activity completion	Students' access
H5P video fruition	Activity completion, xAPI Logstore	Students' access
Quiz evaluation	(Quiz) Statistics	Students' grades
Project evaluation	Grade distribution	Students' grades
Forum participation	Engagement Analytics, Activity completion, xAPI Logstore	Students' level of interaction
Workshop participation	Activity completion, xAPI Logstore	Students' access
Participation trend over time	Engagement Analytics, Logs, Course dedication	Students' overall participation during the course
Survey about the course	Activity completion	Students' opinion and appreciation about the course

As emerges from the table, the data that is suggested to be considered have a twofold value: on the one hand, they allow the teachers to obtain information useful for the evaluation of students and their actual learning; on the other hand, through the same data it is possible to evaluate the effectiveness of the course. Students' assessment in an online course based on a socio-constructivist approach, as anticipated, must take into account several components and cannot be limited to the final numerical evaluation of the acquired knowledge; rather, it must take due to account of the interactive processes and products individually or collaboratively built from the students. Only in this way, it is possible to deduce information on the type of experienced skills and, to some extent, on their acquisition during the course. For this reason, it is suggested first to integrate the results of the final tests with the evaluation of the intermediate tests (quizzes) and the realized products (project management plan). But, for the assessment to correspond to an active learning model, teachers and instructors cannot overlook the role of individual participation, therefore they need to integrate students' grades with data relating to the use of resources and the interaction with one's colleagues over time.

As regards to the evaluation of the effectiveness of the course, all the data considered in the table can offer useful elements when it comes to redesigning our course. For example, educational resources that are highly fruited could indicate particularly attractive lessons, or - on the contrary - complex and difficult-to-understand information. On the other hand, the ones that are poorly consulted could denote materials that are of little interest to students (for example, because they are too elementary). In both cases, teachers should ask themselves what the problem is. In the same way, a quiz reporting high average mean could indicate easy questions and vice versa, as the classical items analysis suggests (Guilford, 1936). Also, analyzing the participation trend over time could give useful insights about external issues to be considered: for instance, if the total amount of students does not participate in the course in a certain period, this could suggest extra-course workload (e.g. exams). Moreover, the interaction level in the course should be considered to understand whether a group is well composed, and the tasks appropriately defined.

Finally, it is just from students' feedback that teachers and instructors can gain crucial data on which to base the course redesign: as final and active users of the planned learning path, they are the most suitable to notify teachers and instructors points of weaknesses and strengths that sometimes the raw data coming from the LMS cannot provide. But, once again, thanks to LA teachers and researchers are enabled to collect, read, and analyze the provided answers and scores.

CONCLUSION

The presented socio-constructivist online course on Engineering Management aimed at promoting the acquisition of domain-related knowledge and nowadays crucial skills. At first, the theoretical foundations of the Learning Scenario were presented, and then how this course can be continuously improved thanks to the Learning Analytics tools embedded in the suggested Learning Management System, which is Moodle.

When applied to education, constructivism calls for a radical redesign of educational goals. Rather than supporting the increase of students' knowledge, the focus should be on the activity the individual can perform in a content domain. Students should not only acquire information, but they should be encouraged to put into practice what they learn. Modern technologies help teachers and instructors to reach this goal, as well as LA data enable them to monitor learning outcomes and improve them, if they adopt approaches and analysis tools consistent with the pedagogical model. The correct interpretation and placement of the quantitative data, as well as a suitable integration of qualitative and quantitative data, is what is required on the one hand by the teachers, on the other by the learning software and the LA techniques, which must necessarily be developed in close connection with the pedagogical assumptions.

Unfortunately, this practice is not yet widespread, as it requires the joint work of several stakeholders. That is, it is necessary that researchers, operators and developers work together around factors such as the development of new tools, the definition of target activities to analyze, and care for ethical aspects related to privacy. In this way, these analytics should become more and more capable of achieving objectives such as: guiding training interventions, providing automatic but personalized feedback, encouraging reflection and interaction in students, and identifying the best practices to follow.

REFERENCES

Biggs, J. B. (2003). *Teaching for quality learning at university* (2nd ed.). Open University Press/Society for Research into Higher Education.

Binkley, M., Erstad, O., Herman, J., Raizen, S., Ripley, M., Miller-Ricci, M., & Rumble, M. (2012). Defining twenty-first century skills. In *Assessment and teaching of 21st century skills* (pp. 17–66). Springer Netherlands. doi:10.1007/978-94-007-2324-5_2

Blumenfeld, P. C., Marx, R. W., Soloway, E., & Krajcik, J. (1996). Learning with Peers: From small group cooperation to Collaborative Communities. *Educational Research*, *25*(8), 37–40. doi:10.3102/0013189X025008037

Brame, C. (2019, November 14). *Flipping the Classroom*. Retrieved January 30, 2020, from https://cft.vanderbilt.edu/guides-sub-pages/flipping-the-classroom/

Branch, R. M. (2009). *Instructional design: The ADDIE approach*. Springer Science & Business Media. doi:10.1007/978-0-387-09506-6

Brown, M., Millichap, N., & Dehoney, J. (2015, April 27). *The Next Generation Digital Learning Environment: A Report on Research*. Retrieved from https://library.educause.edu/resources/2015/4/the-next-generation-digital-learning-environment-a-report-on-research

Clark, R. C., & Mayer, R. E. (2007). *E-Learning and the Science of Instruction: Proven Guidelines for Consumers and Designers of Multimedia Learning*. Pfeiffer.

Delors, J. (1997). Nell'educazione un Tesoro. Roma: Armando.

Design-Based Research Collective. (2003). Design-Based Research: An emerging paradigm for educational inquiry. *Educational Researcher*, *32*(1), 5–8. doi:10.3102/0013189X032001005

Dillenbourg, P. (1999). What do you mean by collaborative learning? In P. Dillenbourg (Ed.), *Collaborative-learning: Cognitive and Computational Approaches* (pp. 1–19). Elsevier.

Dillenbourg, P. (2002). Over-scripting CSCL: The risks of blending collaborative learning with instructional design. In P.A. Kirschner (Ed.), Three worlds of CSCL. Can we support CSCL (pp. 61-91). Heerlen: Open Universiteit Nederland

DiSessa, A., & Minstrell, J. (1998). Cultivating conceptual change with benchmark lessons. In J. Green & S. Goldman (Eds.), *Thinking Practices in Mathematics and Science Learning. Mahwah*. Lawrence Erlbaum Associates.

Educationaldatamining.org. (n.d.). Retrieved from http://educationaldatamining.org/

ENAEE. (n.d.). Retrieved January 30, 2020, from https://www.enaee.eu/

Engeström, Y., & Escalante, V. (1996). Mundane tool or object of affection? The rise and fall of the Postal Buddy. In B. A. Nardi (Ed.), *Context and consciousness: Activity theory and human-computer interaction* (pp. 119–161). The MIT Press.

Gerhana, M. T. C., Mardiyana, M., & Pramudya, I. (2017), The Effectiveness of Project-Based Learning in Trigonometry. *IOP Conf. Series: Journal of Physics: Conf. Series 895*. 10.1088/1742-6596/895/1/012027

Grammatikou, M., Pantazatos, D., Barth, I., Corleto, A., Sansone, N., & Zimniewicz, M. (2019). Pilots Results: Usage And Lesson Learnt Within The First Pilots In Europe. *EDULEARN19 Proceedings*. doi: 10.21125/edulearn.2019.1481

Guilford, J. P. (1936). *Psychometric methods*. McGraw-Hill.

Gustafson, K., & Branch, R. (2002). *Survey of Instructional Development Models* (4th ed.). Eric Publications.

Hernández-García, A., Acquila-Natale, E., & Chaparro-Peláez, J. (2017). Current state of Learning Management Systems' log data-based learning analytics, *17th Biennial EARLI Conference for Research on Learning and Instruction (EARLI 2017).*

Horton, W. (2006). *E-Learning by Design*. Pfeiffer.

Jonassen, D. H. (2006). *Modeling with technology: Mindtools for conceptual change*. Pearson Merrill Prentice Hall.

Lave, J., & Wenger, E. (1991). *Situated learning: legitimate peripheral participation*. Cambridge University Press. doi:10.1017/CBO9780511815355

Le Boterf, G. (1994). De la compétence: Essai sur un attracteur étrange. Les Ed. de l'Organisation.

Ligorio, M. B., Andriessen, J., Baker, M., Knoller, N., & Tateo, L. (2009). *Talking over the computer. Pedagogical scenarios to blend computer and face to face interaction*. Scriptaweb.

Lockyer, L., Heathcote, E., & Dawson, S. (2013). Informing Pedagogical Action. *The American Behavioral Scientist*, *57*(10), 1439–1459. doi:10.1177/0002764213479367

Miao, Y., Sloep, P., Hummel, H., & Koper, R. (2009). Improving the unreliability of competence information: An argumentation to apply information fusion in learning networks. *International Journal of Continuing Engineering Education and Lifelong Learning*, *19*(1), 1560–4624.

Moodle Analytics. (n.d.). Retrieved from https://docs.moodle.org/38/en/Analytics

Moodle Learning Analytics. (n.d.). Retrieved January 30, 2020, from https://docs.moodle.org/33/en/Learning_analytics

Mor, Y., Ferguson, R., & Wasson, B. (2015). Editorial: Learning design, teacher inquiry into student learning and learning analytics: A call for action. *British Journal of Educational Technology*, *46*(2), 221–229. doi:10.1111/bjet.12273

OECD. (2013). *The Survey of Adult Skills: Reader's Companion*. OECD Publishing., doi:10.1787/9789264204027-

Olivé, D. M., Huynh, D. Q., Reynolds, M., Dougiamas, M., & Wiese, D. (2018). A supervised learning framework for learning management systems. *Proceedings of the First International Conference on Data Science, E-Learning and Information Systems - DATA 18*. 10.1145/3279996.3280014

Otero, I., Pantazatos, D., & Docampo, D. (2019). Good Practices Adopting Copyright for Creating and Reusing OERs in NGDLE Environment. *INTED2019 Proceedings*. doi: 10.21125/inted.2019.1359

Otero-González, I., Pantazatos, D., & Docampo, D. (2019). First Steps of Creating an Immersive Scenario for Improving Naval Skills. *ICERI2019 Proceedings*. doi: 10.21125/iceri.2019.0945

Paavola, S., Lipponen, L., & Hakkarainen, K. (2004). Models of innovative knowledge communities and three metaphors of learning. *Review of Educational Research, 74*(4), 557–576. doi:10.3102/00346543074004557

Pantazatos, D., Marinos, C., Giotis, K., Vekris, A., Grammatikou, M., & Maglaris, V. (2018). SeLCont: Synchronized eLearning Content Toolkit. *INTED2018 Proceedings*. doi: 10.21125/inted.2018.2211

Peter, Y., & Vantroys, T. (2005). Platform support for pedagogical scenarios. *Journal of Educational Technology & Society, 8*(3), 122–137.

Rovai, A. (2002). Building Sense of Community at a Distance. *International Review of Research in Open and Distance Learning, 3*(1). Advance online publication. doi:10.19173/irrodl.v3i1.79

Sansone, N., Bortolotti, I., & Buglass, S. (2016). The Trialogical Learning Approach in Practice(s): Reflections from Pedagogical Cases. *Qwerty, Open and Interdisciplinary Journal of Technology, Culture and Education, 11(1).*

Sansone, N., Cesareni, D., Bortolotti, I., & Buglass, S. (2019). Teaching technology-mediated collaborative learning for trainee teachers. *Technology, Pedagogy and Education, 28*(3), 381–394. doi:10.1080/1475939X.2019.1623070

Sansone, N., Cesareni, D., Ligorio, M. B., Bortolotti, I., & Buglass, S. L. (2019). Developing knowledge work skills in a university course. Research Papers in Education. Doi:10.1080/02671522.2019.1677754

Scardamalia, M., & Bereiter, C. (2006). Knowledge Building: Theory, pedagogy, and technology. In K. Sawyer (Ed.), *Cambridge Handbook of the Learning Sciences* (pp. 97–118). Cambridge University Press.

Scardamalia, M., Bransford, J., Kozma, B., & Quellmalz, E. (2012). New assessments and environments for knowledge building. In P. Griffin, B. McGaw, & E. Care (Eds.), *Assessment and teaching of 21st century skills* (pp. 231–300). Springer. doi:10.1007/978-94-007-2324-5_5

Schneider, D. K., Synteta, P., Frété, C., Girardin, S., & Morand, S. (2003). Conception and implementation of rich pedagogical scenarios through collaborative portal sites: clear focus and fuzzy edges. In *Proceedings of the International Conference on Open & Online Learning, ICOOL, International Conference on Open and Online Learning*. University of Mauritius.

Sciarrone, F. (2018). Machine Learning and Learning Analytics: Integrating Data with Learning. *17th International Conference on Information Technology Based Higher Education and Training (ITHET)*. 10.1109/ITHET.2018.8424780

Sergis, S., & Sampson, D. G. (2017). Teaching and Learning Analytics to Support Teacher Inquiry: A Systematic Literature Review. Learning Analytics: Fundaments, Applications, and Trends Studies in Systems, Decision and Control, 25–63. doi:10.1007/978-3-319-52977-6_2

Sfard, A. (1998). On two metaphors for learning and the dangers of choosing just one. *Educational Researcher*, *27*(2), 4–13. doi:10.3102/0013189X027002004

Slavin, R. E. (2011). Instruction Based on Cooperative Learning. In R. E. Mayer & P. A. Alexander (Eds.), *Handbook of Research on Learning and Instruction* (pp. 344–360). Taylor & Francis.

1st International Conference on Learning Analytics and Knowledge (LAK) 2011. (n.d.). Retrieved from https://tekri.athabascau.ca/analytics/

Strijbos, J. W., Martens, R., Jochems, W. M. G., & Broers, N. J. (2004). The effect of functional roles on group efficiency: Using multilevel modeling and content analysis to investigate computer-supported collaboration in small groups. *Small Group Research*, *35*(2), 195–229. doi:10.1177/1046496403260843

Trentin, G. (2003). E-learning come sistema complesso [E-learning as a complex system]. *IJET*, *11*(3), 47–52.

Vygotsky, L. (1978). *Mind in Society* (M. Cole, Trans.). Harvard University Press.

Wartfoskij, M. (1973). *Models*. D. Reidel.

Watzlawick, P., Beavin, J. H., & Jackson, D. D. (1971). *Pragmatica della comunicazione umana*. Astrolabio.

Webb, N. (1985). *Student interaction and learning in small groups: A research summary*. Learning to Cooperate, Cooperating to Learn. doi:10.1007/978-1-4899-3650-9_6

Weinberger, A. (2008). *CSCL scripts: Effects of social and epistemic scripts on computer-supported collaborative learning*. VDM.

Wenger, E., Trayner, B., & De Laat, M. (2011). *Promoting and assessing value creation in communities and networks: A conceptual framework*. Ruud de Moor Centrum. Available at http://www.knowledgearchitecture.com/downloads/Wenger_Trayner_DeLaat_Value_creation.pdf

What is xAPI aka the Experience API or Tin Can API. (2020). Retrieved 30 January 2020, from https://xapi.com/overview

Xu, M., David, J. M., & Kim, S. H. (2018). The Fourth Industrial Revolution: Opportunities and Challenges. *International Journal of Financial Research*, *9*(2), 90. doi:10.5430/ijfr.v9n2p90

ADDITIONAL READING

H5P. (n.d.). Retrieved February 4, 2020, from https://h5p.org/

Introduction to SeLCont. (n.d.). Retrieved February 4, 2020, from http://www.netmode.ntua.gr/main/index.php?option=com_content&view=article&id=142&Itemid=93

Jitsi Meet. (n.d.). Retrieved February 4, 2020, from https://jitsi.org/

Moodle - Open-source learning platform. (n.d.). Retrieved February 4, 2020, from https://moodle.org/?lang=en

Studio, O. B. S. (n.d.). Retrieved February 4, 2020, from https://obsproject.com/

xAPI.com Homepage: What is xAPI (the Experience API). (n.d.). Retrieved January 31, 2020, from https://xapi.com/

KEY TERMS AND DEFINITIONS

Educational Data Mining: A research field concerned with the application of data mining, machine learning and statistics to information generated from educational settings.

Learning Analytics: The measurement, collection, analysis, and reporting of data about learners and their contexts, for purposes of understanding and optimizing learning and the environments in which it occurs.

Learning Management System (LMS): A software application for the administration, documentation, tracking, reporting, and delivery of educational courses, training programs, or learning and development programs.

Public License: A license by which a copyright holder as licensor can grant additional copyright permissions to all persons in the general public as licensees.

Shareable Content Object Reference Model (SCORM): A collection of standards and specifications for web-based electronic educational technology (also called e-learning).

Chapter 4
Innovative Learning and Education Practices in European Universities:
A Pathway to Modern Pedagogy

Marcos Komodromos
https://orcid.org/0000-0002-2910-6541
University of Nicosia, Cyprus

EXECUTIVE SUMMARY

In the dynamic world that is associated with unpredictable informatization and globalization, it is indispensable for education to be sustainable and continuously evolve to facilitate in fulfilling the learners' needs in the current digital technologies and knowledge era. In this chapter, the author discussed how innovative learning and education practices for modern pedagogy have been developed in the European universities system. Specifically, the author provides a delineation of the attributes, pedagogy, ideology, ethical dimensions, and practical applications of innovative learning education practices. In the chapter, the author identified the different practices such as massive open online courses (MOOCs), blended learning, and flipped classrooms that have been applied at different European universities. Thus, the information discussed in this chapter is relevant to university educators and electronic learning practitioners.

DOI: 10.4018/978-1-7998-4063-3.ch004

INTRODUCTION

Preparing graduates to resolve the emerging contemporary challenges that are caused by change requires consistent review and update of the pedagogies used by teachers (Peterson, 2018). The 21st century is characterized by the proliferation of scientific and informative technology, the creation of knowledge, and the advancement of research and education (Mykhailyshyn et al., 2018). Burke et al. (2017) supported that the attributes of the modern society, especially, the increased innovation have resulted in the adoption and use of technology in learning. The social changes that have been caused by Web 2.0 tools have had a significant impact on the learning practices for modern pedagogy. Bennett et al. (2012); Kovarova and Simkova (2014); Yadav and Patwardhan (2016); and Henderson et al. (2015) posit that the emergence of educational innovation such as electronic learning, interactive multimedia, mobile learning, wireless devices, and hypermedia over the past 10 years has helped in promoting pedagogy at institutions of higher learning.

In the 21st century, universities in Europe and other parts of the world lost the monopoly of producing and disseminating knowledge as a result of the introduction of innovative learning and education practices (Freitas & Paredes, 2018). The challenges European universities are experiencing in pursuit of fulfilling the demands of the society can be categorized into economical, technological, and faculty-based issues. First, higher learning institutions models are based on the economy (Mehaffy, 2012). The model is focused on educating more students and promoting learning at a lower cost. Second, education at universities is affected by technological advancements because the Information Communication Technology (ICT) requires changes to the teaching and learning process. Third, the faculty is required to enroll in professional development activities that impart them with the competencies to offer the quality of education that is demanded in society. The faculty members at universities are required to be innovative and apply current academic teaching and learning methods. Hence, learning at universities should be innovative, engage students, and fulfill the learners' educational needs (Freitas & Paredes, 2018; Mehaffy, 2012; & Norgard et al., 2018).

In the modern era, the foundation roles of education include (a) guarding the quality of knowledge produced and disseminated, (b) promoting reflective and critical thinking, and (c) structuring learning (McGrath et al., 2016). Institutions of higher learning are the core validators of scientific theories, which is done through falsification procedures such as peer-reviews. The introduction of ICT has resulted in changes in academic research and publication because of the prevalent reliance on open science to disseminate findings. Faculty members in universities are responsible for promoting reflective and critical thinking by imparting students with competencies and skills (Brennan et al., 2014). In universities, structured learning is done to help graduates transition to the labor market (McGrath et al., 2016).

This chapter was written to discuss the importance of technology in higher education and the practical applications of innovation education in European universities, specifically, MOOC, blended learning, and flipped blackboard. The authors began the chapter by making the reader aware of the factors that have supported the adoption of innovative learning and education practices in European universities, which have significantly affected the institutions' monopoly of creating and disseminating knowledge. The authors then proceed to provide background information on the factors that have influenced the adoption of innovative learning and educational practices in European universities.

In the background section, the authors focused on the impact of innovative education and learning practices in Europe and argued for the importance of the university management and governance to provide adequate resources. A discussion of the characteristics of innovative education is relevant to this chapter because it helps the readers understand that not all technology that is applied in learning can be categorized as innovative. The background section supports the readers' ability to determine the technology the can be categorized as innovative learning and educational practices. Similarly, a discussion on the ideology of innovative education in Europe was included to enable the readers to understand that the adoption of innovative technology in universities does automatically result in a change in pedagogy. The section is relevant because it contains a discussion on the concepts that support the ideology of innovative learning and educational practices, which could result in the advancement of modern pedagogy.

The author equally emphasized the importance of innovative learning education and the different practices that have been adopted in Europe. The chapter can be improved by adding a discussion on the issues that hinder the adoption of innovative learning and educational practices in specific European universities. Conversely, there are limited peer-reviewed studies assessing the specific innovative technologies in European universities. A weakness of the chapter is that although different types of MOOCs, blended learning, and flipped classrooms and their limitations were discussed, the limited availability of literature hindered the comprehensive discussion on the innovative learning and educational practices. Another weakness is that the chapter over-emphasizes on MOOCs because the majority of the existing literature focuses on the innovative practice compared to others such as blended learning, and flipped classrooms. The weaknesses limit the comprehensive analysis of the practical application of innovative education in Europe. The section can be improved by including examples of the European universities that have adopted innovative practices and the manner they have fostered modern pedagogy.

The first strength of the chapter is that it provides the reader with a comprehensive overview of the concept of innovative learning and educational practices. A second strength is that the augments in the chapter are based on findings from current and credible literature. A third strength is that the chapter provides academicians, and professionals in the field of engineering management education with a discussion of the limitations associated with the MOOC, blended learning, and flipped classroom practices implemented in European universities. The strengths can be applied to enhance the limited analysis of the issues because the chapter is based on a synthesis of the manner the concept of innovation and technology has impacted pedagogy in European universities. Additionally, a discussion on the ethical issues that are significant barriers to the adoption of innovative learning and education practices in Europe was included in this chapter.

Although innovative learning and education practices have been adopted in European universities, they are associated with limitations, which hinder efficacy. Consequently, in the solutions and recommendations section of this chapter, the author explain the different practices that can be adopted by university educators and/or e-learning practitioners to mitigate the existing barriers and foster future efficacy and development. The authors concluded by providing future researchers with recommendations. This chapter is properly directed towards the targeted researches, academics, and professionals in the field of engineering management education, specifically those in leadership and management positions. In the chapter, it was identified that the implemented innovative learning and educational practices in European universities are associated with limitations. This chapter provides the target audience with an understanding of the issues and recommendations that are based on existing literature.

BACKGROUND

Traditionally education has been geographically limited and spatially fixed (Balkin & Sonnevend, 2016). The fixed and geographically limited nature of education requires learners to travel to a specific location and reside within or near the institution. The infrastructure and human resources available determine the number of students enrolled at universities. In the modern era that is characterized by increased student population and demand for higher learning, there is a need for pedagogical changes and implementation of technology-enhanced learning (Balkin & Sonnevend, 2016 & Timmis et al., 2015). The advent of innovative technology has resulted in the digitization of teaching and learning resources that have increased the replicability and propensity content (Pearce et al., 2010). The availability of academic content online has facilitated in decreasing the limitations imposed by the physical and human resources in a learning institution. The adoption of innovative learning, technology-enhanced education, and web tools are associated with increased access to knowledge, cost reduction, and improved pedagogy (Yadav & Patwardhan, 2016).

In the mid-1980s, Open Education Resources (OERs) were the first initiatives towards supporting the concept that knowledge should be disseminated freely through the internet (Kanwar & Uvalic-Trumbic, 2015). Since the 1990s, there has been a significant increase in the adoption of innovative learning practices in higher education as a technique for supporting efficient learning (Kirkwood & Price, 2013). Efficient learning should encourage reflection, enable dialogue, promote collaboration, foster creativity, motivate learners, and support the application of the theories learned into practice (Conole, 2016). Over the years, there has been an increase in the adoption of innovative learning and educational practices. The application of innovative learning and educational practices is associated with positive outcomes. Henderson et al. (2015), in their study, supported the use of innovative learning in promoting the students' outcomes.

INNOVATIVE EDUCATION AND LEARNING PRACTICES IN EUROPE

The leaders and managers at institutions of higher learning experience the challenge of increasing the spread of knowledge and accommodating more learners at a lower cost (Daniel et al., 2015). A possible solution to the dilemma experienced at universities in the adoption of innovative education and learning practices that have an impact on the teaching methods, programs, and curricula (McGrath et al., 2016). The adoption of innovative education and learning practices is congruent with the European Union 2020 strategy for creating effective governance mechanisms that promote excellence (European Commission [EC], 2010). Innovative education and learning practices such as MOOC have facilitated in transforming education at institutions of higher learning (Freitas & Paredes, 2018). The implementation of innovative learning techniques supports the universities' aims of educating adults, promoting competence, and facilitating continuous training. Innovative learning practices support face-to-face teaching, promote both open and flexible learning, and foster the creation of hybrid models that can be applied in teaching to increase the learners' experience (Bryant, 2018; Daniel et al., 2015; & Yadav & Patwardhan, 2016).

In the context of Innovative Education (IE) learning strategies and teaching in modern pedagogy, the seven stages that educators must adhere to promote technology adoption include awareness, informational, personal, management, consequences, collaboration, and reinforcing (Hall et al., 2013). At the awareness stage, a professor has little concern and involvement with technology and innovation. Educators at the informational stage have a general awareness of technology and innovation and are interested to learn and practice using IE techniques. Professors at the personal stage lack a comprehensive understanding of the technological and innovation demands. Educators at the management stage are those studying about the different processes, resources, and tasks of using technology and innovation in modern pedagogy. At the consequence stage, a professor focuses on the manner technology and innovation can impact the students. Differently, the professors who are at the collaborative stage focus on coordinating with others when using technology and innovation. At the reforming stage, a professor focuses on the exploration of more universal benefits from technology and innovation (Hall et al., 2013).

When professors apply digital technologies during lectures, they can effectively enhance pedagogical practice and student learning (Schuck & Kearney, 2008). Several studies reveal that learning experiences are facilitated by social interactions in the classroom between professors and learners through discussions, negotiations, and workshops (Cogill, 2010). As a result, after professors become familiar and competent users of technology, they may be ready to create new pedagogical models for their courses to enhance learning. Freitas and Paredes (2018) conducted a study to assess the manner experiences among faculty members affect innovative practices. Also, the researches wanted to understand the correlation between ICT innovation and the faculty members' roles. Data were collected through a case study involving faculty members who had used innovative learning and education practices. The researchers found that the faculty were of the opinion that adopted practices would promote collaborative and individualized learning. Freitas and Paredes (2018) concluded by recommending that the university management and governance should provide the required resources to facilitate the adoption of innovative learning and educational practices, which augment modern pedagogy.

Characteristics of Innovative Education in Europe

The introduction of innovative learning and educational practices has facilitated in increasing the competitiveness and the quality of higher education in Europe (Lasakova et al., 2017). Innovative education supports adaptive learning techniques that promote the personalization of the content presented to the students (Daniel et al., 2019). The characteristic of innovative education is that knowledge is disseminated at massive scale, it facilitates the interconnection of students from different nations and cultures, fosters contemplation and reflection, augments traditional pedagogical practices at universities, and supports embodied and personalized learning (Daniel et al., 2019; Murugesan, 2019; & Yadav & Patwardhan, 2016).

The introduced innovative and educational practices should be pedagogically effective, enhance the students' experience, and promote quality assurance (Conole, 2016). Innovative education practices such as MOOCs facilitate the dissemination of curated and quality resources and information to numerous learners around the globe (Ferguson & Sharples, 2014). The ability of innovative education practices to support the provision of knowledge to numerous individuals through the use of a limited number of trained educators results in economic benefits. The massive pedagogy is associated with advantages such as collaboration, co-construction of knowledge, and cooperation (Ferguson & Sharples, 2014). Innovative education is associated with the attributes of contemplation and reflection because the pedagogy is based on the crossover, argumentation, and incidental learning (Kalyani & Rajasekaran, 2018 & Murugesan, 2019).

Innovative learning and education practices can be competency, work or employment, outcome, inquiry, and/or student-based, (McGrath et al., 2016). In reference to the attributes of innovative education supporting traditional pedagogical practices at institutions of higher learning and fosters personalized learning are supported by the interactive nature of the practices (Daniel et al., 2019; Murugesan, 2019; & Yadav & Patwardhan, 2016). Innovative education tools facilitate scholar-centered approaches and participation that impart the students with critical evaluation and communication skills, which make the learners marketable employees. Also, innovative education is associated with enhanced student engagement and motivation (Murugesan, 2019).

Ideology of Innovative Education in Europe

The process of innovative education and learning involves the application of technology to foster the use and creation of knowledge (Pisanu, 2014). Conversely, the introduction of technology in institutions of higher learning does not result in an automatic change of pedagogy. The ideology of innovative education is hindered when technology is applied to recreate traditional teaching practices. There is a need for higher education systems to be based on effective strategies and models that help in leveraging pedagogical innovation (Norgard et al., 2018). Innovative education and learning practices support the intensification of the educational process, increases access to knowledge, promotes efficiency, and imparts learners with additional skills that cannot be provided through the traditional techniques (Balkin & Sonnevend, 2016). The ideology of innovative education is based on crossover, argumentation, and incidental learning, computational thinking, and understanding through doing science (Kalyani & Rajasekaran, 2018; Murugesan, 2019).

Crossover Learning

Crossover learning involves linking formal and informal settings to promote a comprehensive understanding and motivate learning (Kalyani & Rajasekaran, 2018; Murugesan, 2019). The innovative education method involves incorporating experiences from everyday life as a technique for providing students with authentic and engaging opportunities to increase their interest. The adoption of crossover learning techniques requires educators to make changes in the manner curriculums are planned, and students are assessed (Kalyani & Rajasekaran, 2018; Murugesan, 2019).

Learning Through Argumentation

This ideology of innovative learning involves advancing the learners' understanding of concepts in science and mathematics through arguing (Kalyani & Rajasekaran, 2018 & Murugesan, 2019). In the process of proposing, critiquing, and defending, the students understand different concepts and comprehend the manner that scientists collaborate to formulate or refute ideas. The ideology of learning through argumentation enables learners to argue scientifically, listen actively to peers, and provide constructive criticism assessed (Kalyani & Rajasekaran, 2018 & Murugesan, 2019).

Incidental Learning

Incidental learning is unplanned and happens when a student is performing seemingly unrelated activates to learning (Kalyani & Rajasekaran, 2018). In this innovative educational method, educators and a structured curriculum are not required. Thus, incidental learning does not result in certification. Incidental learning supports reflective learning among learners (Kalyani & Rajasekaran, 2018).

Computational Thinking

Computational thinking promotes problem-solving and involves breaking problems into manageable portions, recognizing patterns, eliminating unimportant concepts, and developing the techniques required to formulate solutions (Murugesan, 2019). Computational thinking allows learners to structure problems and formulate solutions effectively. The purpose of innovative education that is based on computational thinking is to enable students to apply their skills in resolving complex challenges in different disciplines (Murugesan, 2019).

Learning by Doing Science

The process involves engaging learners with scientific equipment as a technique for improving understanding and increasing motivation (Kalyani & Rajasekaran, 2018). Learning by doing science helps decrease barriers and complements the concepts covered in the textbooks. The innovative practice provides students with a real scientific experiment in fields such as but not limited to, biology, engineering, hydraulics, and robotics (Kalyani & Rajasekaran, 2018).

Pedagogy of Innovative Education in Europe

It is indispensable for higher education to help students gain competency-oriented skills and actively participate in the learning process. It is difficult to transform pedagogy, but the introduction of adaptive and digital learning programs, collaborative technologies, and online resources have supported changes in higher education practices (Uskov et al., 2018). Initially, the effective implementation of strategic pedagogical changes was limited by the notion that technology would hinder the efficacy of teaching (Pearce et al., 2010). The resistance to technological change in education in the past has limited the growth of online and blended learning. However, pedagogical change that is based on technology is essential in ensuring that the education provided in institutions of higher learning is congruent with the market demands. Technology based pedagogical change supports the enhancement of practice and fosters innovation (Bryant, 2018).

The second generation of internet technologies that are commonly referred to as Web 2.0 are innovative pedagogical tools that have a significant impact on research and education (Yadav & Patwardhan, 2016). The concept of Web 2.0 has supported the introduction of e-learning 2.0 and education 2.0. The interactive nature of the tools has underpinned the adoption of web-based teaching and research in higher education. Educators have supported that the incorporation of web tools in higher education learning helps in preparing students to achieve the job requirements of the future workplace. Also, Web tools provide the students and instructors with a platform for creating their content and sharing it in the forms of wikis, blogs, and social bookmarking (Yadav & Patwardhan, 2016).

Cogill (2010) argued that social interactions such as discussions, negotiations, and workshops that occur between the professors and students promote learning experiences. The growth of modern civilization is determined by the quality and accessibility of education (Burke et al., 2017). Thus, the need for new forms of educational adaptation is supported by the increasing demand for active learning. Higher-education policies should support the principles of educational, scientific, innovation, and technological activities in tertiary learning institutions. Innovation in institutions of higher learning is a core component of pedagogical changes (Burke et al., 2017).

Practical Applications of Innovation Education

The emergence of new disruptive technology has significantly affected education and learning (Conole, 2016 & Kop et al., 2011). The introduction of the European Higher Education Area (EHEA) that was developed as a result of the Bologna Declaration has supported modification in tertiary education in Europe (Olivan et al., 2019). The EHEA has supported the change from a teaching-learning process to a student-centered approach. In response to the need for the adoption of active learning techniques, universities adopted new models that are based on innovation. Some of the innovative learning and education practices adopted in European universities include MOOCs, blended learning, and flipped classrooms (Adair et al., 2014; Garcia et al., 2015; Kop et al., 2011; & Mehaffy, 2012).

Massive Open Online Courses

The MOOCs are a collection of learning content that students can access via the web (Adair et al., 2014; Garcia et al., 2015; & Kop et al., 2011). The innovative learning and education practice provide students with access to quality materials on different platforms such as blogs, social media, forums, and online portals. Barak et al. (2016) defined MOOCs as a platform that provides learners with free access to education. Examples of MOOCs in Europe are FutureLearn, OpenHPI, and Leuphana (Brennan et al., 2014). FutureLearn and OpenHPI are meant to attract numerous students from around the world and provide students with access to quality courses. Leuphana is a German-based MOOC that provides overseas students with opportunities to demonstrate their intellectual capabilities before migrating. The program also supports lifelong learning opportunities (Brennan et al., 2014).

MOOCs augment education models and should be applied to facilitate the application of collaborative methods to improve the students' knowledge and enhance learning (Freitas & Paredes, 2018). The application of MOOC provides learners with flexible and free courses that enable universities to disseminate education to a diverse and broad audience (Daniel et al., 2015 & Freitas & Paredes, 2018). The dimensions of MOOCs are massiveness and openness (Conole, 2016; Pilli & Admiraal, 2016). Massiveness refers to scale, while openness is the lack of place, pace, accessibility, adaptability, and cost-related barriers in enrolling for the course. In their study, Pilli and Admiraal (2016) categorized MOOCs into four dimensions.

Small Scale and Less Open MOOCs

Small scale and less open MOOCs include Small Private Online Courses (SPOCs), task-based MOOCs, and group MOOC (Pilli & Admiraal, 2016). The MOOCs have 200 to 500 participants and are less open because a fee is required to access some content. They include SPOCs, group MOOCs, and task-based MOOCs. SPOCs are similar to flipped classrooms. In SPOCs, students collaborate with the lectures through online or face-to-face modes. The limited number of students allows educators to provide specialized learning. In group MOOCs, the participants have a mentor, and the learners are encouraged to collaborate to decrease dropout rates. Learners are selected through the use of a software that considers the participants' skills, knowledge, and socio-demographic background. Task-based MOOCs are a blend of instructivism and constructivism focused on imparting the learners with skills to complete specific activities (Pilli & Admiraal, 2016).

Small Scale and More Open MOOCs

The course materials and exams in this type of MOOCs are freely accessible to all participants. Also, learners are encouraged to share their outputs through common creative licenses (Pilli & Admiraal, 2016). The licenses allow students to gain copyright to work created and allows other learners to copy, disseminate, and use the content for non-commercial purposes. The MOOCs include connectivist MOOCs (cMOOCs), Big Open Online Course (BOOCs), Distributed Online Collaborative Courses (DOCCs), Participatory Open Online Courses (POOCs), Little Open Online Courses (LOOCs), Game-based MOOCs (gMOOCs), Project-based MOOCs (pMOOCs), adaptive MOOCs and network-based MOOCs (Pilli & Admiraal, 2016).

cMOOCs are based on a constructivist approach, while xMOOCs incorporate traditional and behaviorist perspectives. In cMOOC, students are educated on the content that they need to learn. cMOOCs are based on four core activities that include feedforward, aggregation, re-blending, and repurposing of content (Kop et al., 2011). Aggregation refers to the activity of accessing various resources to read, watch, or play. Re-mixing or re-blending involves the process of creating content based on information read or watched. Repurposing is an activity that supports learners to create content on their own. Feedforward involves learners sharing their work with peers in the course (Kop et al., 2011). cMOOCs are suitable for social learning. In cMOOCs, there are concerns about the level of support provided by educators and the extent that the model allows peer-to-peer feedback as a technique for promoting cognizance (Adair et al., 2014; Kop et al., 2011).

BOOCs are open to a limited number of approximately a maximum of 500 learners, which promotes interactions (Pilli & Admiraal, 2016). DOCCs are focused on promoting collaborative and institutional activities. In practice, courses are based on learning resources. POOCs are developed to promote collaboration among the participants. LOOCs are different from most MOOCs because they contain a standardized assessment and grading feature. In LOOCs, learners are required to participate in workshops and write essays. Presque Isle's by OpenU is an example of a LOOCs. In gMOOCs, which are based on connectivist and decentralized concepts, digital games are used to facilitate learning and group problem-solving. In pMOOCs, learners collectively develop small innovative projects and are suitable for media education. Adaptive MOOCs are structured and sequential, which limits personalized learning. Network-based MOOCs are based on the connectivist-style and are used to develop knowledge through social conversations.

Large Scale and Less Open MOOCs

These MOOCs have limitless participants but restricted access to content (Pilli & Admiraal, 2016). Openness is limited through registration fees, course design, target groups, and customized content. Examples of large scale and less open MOOCs include Vocational Open Online Courses (VOOC), Hybrid Open Online Course (HOOC), miniMOOC, and Personalized Open Online Course (POOC). In VOOC, pedagogies are based on vocational learning, and real-world examples are used as the basis of procedures and tasks. HOOCs are a version of flipped classrooms that are used to enhance collaboration among participants. In mini MOOCs, fewer skills and content are covered, and they last for hours or days. POOCs are based on non-standardized content, and education is adjusted based on the users.

Large Scale and More Open MOOCs

In these MOOCs, knowledge, and ideas are disseminated by educators through course content and videos (Pilli & Admiraal, 2016). The students can determine the style and time of the study, but flexibility is limited in terms of course objective, content, assessment, and sequencing. Examples of large scale and more open MOOCs include xMOOC, transfer MOOC, made MOOC, async MOOC, Self-Paced Online Course (SPOC), Content-based MOOC, Flex-MOOC, and Massive Online Research (MOOR). xMOOC supports learning in large and structured classes. Feedback in xMOOC is minimal, and it is usually provided on a peer-evaluation basis. xMOOCs facilitate in creating a codified, structured, and traditional educational experiences that are similar to the traditional programs (Adair et al., 2014; Pilli & Admiraal, 2016). Conversely, different from traditional courses, xMOOCs are supported by an infrastructural mechanism that enables the innovative learning and educational practice to serve thousands of learners (Adair et al., 2014).

Transfer MOOCs are courses designed to be taught in lecture rooms, but they are uploaded on a MOOC platform and are effective when used to transfer knowledge (Pilli & Admiraal, 2016). Differently, made MOOCs involve the use of videos and interactive content that facilitates not only transferring knowledge but also promoting problem-solving. Asynch MOOCs are open and don't have fixed start or end dates. The flexibility nature of the MOOCs decreases the prevalence of dropouts. SPOC allows students to schedule their learning based on preference. Educators teach Content-based MOOCs from major universities, and evaluation is done through automatic methods. Flex MOOCs allow students to learn content that is congruent with their strengths, preference, and needs. The flexibility of the MOOC will enable students to choose their learning objectives, assignments, content, and assessment modes. MOOCs are research-focused and allow individuals around the world to collaborate to translate theory into practice and receive mentorship from experts in the respective fields (Pilli & Admiraal, 2016).

Effectiveness and Limitations of MOOCs

Bralic and Divjak (2018) conducted a qualitative study to determine whether MOOC facilitates in creating a positive learning experience for the students and helps part-time learners achieve the required outcomes. Also, the researchers assessed the challenges experienced by learners when using MOOCs and the students' motivation to learn. The researcher found that MOOCs support the learners' experience and enable part-time students to achieve the required outcomes. In the study, language was identified as a core barrier for students when using MOOCs. It was also identified that MOOCs facilitates in motivating students to learn. Bralic and Divjak (2018) concluded that there are benefits for integrating MOOC with lecture-based teaching.

In another study, Lowenthal and Hodges (2015), assessed the quality of MOOCs. In the study, the researchers argued that individuals who critique MOOCs indicate that technology could result in exam cheating and replace teachers. Also, some individuals are skeptical about the pedagogical approach of MOOCs. Lowenthal and Hodges (2015) found that although not all, MOOCs are effectively designed and support learning. The researchers concluded that MOOCs support innovative learning and education practices. Similar to Lowenthal and Hodges (2015), Margaryan et al. (2015) conducted a study to assess the quality of MOOCs. Margaryan et al. (2015) analyzed 76 randomly selected MOOCs and found that although the innovative learning and education practices are well-developed, their instructional design quality is limited.

The first significant limitation of MOOCs is the prevalent student dropout rate (Daniel et al., 2015). Approximately 60% to 90% of students who enroll from MOOCs dropout. Second, European universities are publicly funded, and there exists a debate about whether the institutions have a right to develop MOOCs because the cost of developing the practices is significant. The process requires personal time commitment and effort. Third, in European universities, there is an issue in the manner teachers, professors, and tutors should be remunerated. Fourth, there are challenges in how the educators' participation in MOOCs should be integrated with their roles, especially with the limited accreditation of the practices (Daniel et al., 2015).

Blended Learning

The increased access to the internet and support of the efficacy of online learning in promoting the role of universities has resulted in the adoption of blended courses. Blended courses involve combining traditional and modern strategies for learning and education (Mehaffy, 2012). The practice involves a combination of both face-to-face and online learning. Blended courses allow faculty members to use the internet to deliver instructions, offer students with multiple learning plaforms, and allows educators to spend more time on the course content. Also, blended learning enables educators to focus their time on teaching and creating learning programs that align with the students' needs (George-Walker & Keeffe, 2010). Pisoni (2019) indicated that blended learning allows educators to leverage the internet in delivering optimal education. The approaches of blended learning can be categorized as low, medium, and high (Alammary et al., 2014).

A low-impact blend involves adding more online-based activities into an already existing course. In the medium-impact blend approach, activities of an existing course are replaced with new ones. The approach involves using the Learning Management System (LMS) to facilitate online discussions. The high-impact blend approach involves creating a new course, and the process allows the integration of the advantages associated with online and face-to-face methods (Alammary et al., 2014).

An example of innovative learning and education practices are those adopted at a German-based institution of higher learning: Bavaria Virtual University (BVU). At BVU, there exist free and low fee personalized online courses that are developed according to blended learning (Brennan et al., 2014). Anglia Ruskin University has Degrees at Work (DaW) program that provides individuals in the workplace with an opportunity to earn their diplomas and degrees (McGrath et al., 2016). Blended courses are among the different forms of innovative learning, and educational practices are applied to facilitate in delivering the DaW program. Learners have online and telephone access to tutors and a platform to consult with peers. The students are required to spend 8 to 10 hours learning and subsequently apply the content at work. The program enables students to develop new skills and knowledge that they apply in their professions (McGrath et al., 2016).

Effectiveness and Limitations of Blended Learning

Innovative learning education practices can be applied to provide students with the opportunity to work together collaboratively (Bryant, 2018). George-Walker and Keeffe (2010) conducted a case study to assess the impact of blended learning on the students. The researchers found that blended learning supports student engagement and helps in meeting the learners' needs. George-Walker and Keeffe (2010) recommended that blended courses should support self-reflection and encourage personalized learning. Medina (2018) analyzed the blended learning implemented in Open University, the University of Trento, the Universidad Nacional de Educación a Distancia, the University of Southampton, and the University of Cambridge. In the study, the researcher incited that blended learning supports flexibility and personalized learning (Medina, 2018).

Additionally, Kintu et al. (2017), in their survey of 238 participants, found that blended learning facilitates in increasing the learners' analytical skills and ability to assess knowledge from different sources critically. Although the adoption of blended learning positively impacts the manner education is disseminated, the effectiveness of the innovative practice is limited by various disadvantages. Some of the limitations include log-in problems, user-unfriendly interfaces, decreased motivation to learn, and reduced creativity in searching for knowledge (Szadziewska & Kujawski, 2017).

Flipped Classrooms

Flipped classrooms are a form of blended learning (Nat, 2015). The innovative practice involves blending classroom instruction with online systems, which allows collaboration and interactions among the students and teachers. In flipped courses, the content is delivered as homework, and during the lectures, learners collaborate to increase understanding and decrease misconceptions (Mehaffy, 2012). The flipped model is an innovative learning and education practice that ensure learning environments are focused on comprehension and interaction. In 1990, Eric Mazur, a physics professor at Harvard, was among the early adopters of the flipped courses (Mehaffy, 2012). The concept was later adopted by the founders of Khan Academy (Stohr & Adawi, 2018).

In the flipped or inverted classroom model, the students are encouraged to watch video lectures as a preliminary activity before attending the class (Stohr & Adawi, 2018). Flipped classrooms are based on constructivist and social-constructivist perspectives on learning, which facilitate in increasing active learning and the importance of scaffolding. Thus, the attributes of flipped classrooms are that a significant amount of the content is provided to students outside the classroom; the learners actively interact with the materials, peers, and instructors; and the scholars are required to complete some activities preliminary to the actual class (Simonson, 2017).

At the Queen Mary University of London, an innovative learning and education practice referred to as *lecture capture* software, has been implemented (McGrath et al., 2016). The software allows educators to record and upload live lectures and record materials using a Q-Review system. The e-learning unit manages the Q-Review system at the institution. The e-learning unit at the university evaluated to assess the effectiveness of the *lecture capture* and found that the practice is useful for both teachers and students. A limitation associated with *lecture capture* is the static nature of the cameras, which means that the lectures must stand in the same position. Conversely, *lecture capture* is useful, especially for students living with disabilities or experiencing learning difficulties (McGrath et al., 2016).

Stockholm school of economics in Riga has an innovative learning and education practice that is in the form of an interactive game on entrepreneurship (McGrath et al., 2016). The virtual game enables students to establish a company through simulation. The innovative technology provides feedback on the remaining cash balance after every decision. During class, the lectures analyze the learners' results, and the process facilitates in imparting the students with an understanding of the initial company development stages. The strategies support the implementation of the innovative learning. Education practice at Stockholm school of economics provide the educators with the freedom to decide the program content and teaching methods (McGrath et al., 2016).

Effectiveness and Limitations of Flipped Classrooms

In their study, Mortensen and Nicholson (2015) found that flipped classrooms facilitate in improving the students' performance and critical thinking because students have active roles that are guided by educators. Similarly, Bergfjord and Heggernes (2016); Heijstra and Siguroardottir (2017); O'Flaherty and Phillips (2015); Olivan et al. (2019); and Wang & Zhu (2019), supported that flipped classrooms are associated with improved academic performance, enhanced staff satisfaction, and facilitate learning. In another study, Nat (2015) reported that the adoption of the flipped classroom model results in increased engagement among educators and students in universities.

Conversely, the use of flipped learning among first-year students results in decreased performance. Roehling et al. (2017) argued that avoiding destructions while simultaneously taking notes and watching videos could be a challenging process. Other identified limitations of flipped classrooms are decreased motivation among students to learn and technical problems such as user-unfriendly interfaces and issues downloading or viewing the content (Aydın & Demirer, 2016 & Roehling et al., 2017).

Ethical Dimensions of Innovative Education in Europe

Higher education is affected by innovations that occur around the world (Lasakova et al., 2017). Compliance with innovation has resulted in a need for the introduction of new educational models that are suitable for the population. The introduced educational models support distance learning, promote the ability of students to collaborate, increase academic performance, and foster teacher-learner interactions (Balkin & Sonnevend, 2016). Conversely, the adoption of innovative education and learning institutions has resulted in the occurrence of ethical challenges. The ethical dimensions of innovative education are related to confidentiality, identity, surveillance, plagiarism, and cheating (Balkin & Sonnevend, 2016).

Innovative learning and education platforms provide the students with the ability to effectively search and access multivariate data (Balkin & Sonnevend, 2016). The interactive and exploratory nature of innovative educational technologies allows users to assess the scope of their understanding. Conversely, the course presented could result in profound variance in the students' understanding of the concept. The unguided expose learners to course content could affect the role of teachers.

The first ethical issue is associated with the intellectual property rights. Students can infringe content protected by patent, trade secret laws, and copyright by cutting or copying and pasting materials in their work, which leads to academic fraud (Balkin & Sonnevend, 2016). The second issue is privacy because there lacks a distinction between private and public data when innovative education incorporates social media (Balkin & Sonnevend, 2016). In online education, a significant number of privacy issues are associated with big data. Big data supports the permanent storage of the data placed on the internet. A significant number of students do not understand the implication of the information shared, and this creates a social-ethical challenge.

To some extent, the adoption of social media as an innovative learning and educational practice is hindered by privacy concerns. The third ethical issue is associated with the accuracy of the information developed. Innovative education practices support the rapid dissemination of information. The disseminated information could be false, misleading, incomplete, and unreliable, which affects the concept of education at its core (Balkin & Sonnevend, 2016).

SOLUTIONS AND RECOMMENDATIONS

In the increasingly dynamic student population, the higher education system should be learner-centered (Yadav & Patwardhan, 2016). The management and educators should be focused on promoting creativity, efficiency, and innovation through the introduction of effective teaching and learning models. Conversely, addressing the 21st-century challenges in higher education requires more than just an introduction of new course models and technology. The transformation of learning is complex, which creates a need for educators to reconsider the concepts of teaching and learning. In their studies, Norgard et al. (2018) and Pisanu (2014) support that the effective implementation of innovative learning and education practices can result in pedagogical changes and improved learning outcomes.

The outcomes of innovative learning and education include increased culture of change and innovation, rise in cross intuitional collaboration, proliferation in blended learning, growth in open education resources, and the emergence of the need to redesign learning spaces. However, the adoption of innovative learning and education practices is associated with pedagogical, assessment, and related ethical limitations. This section contains solutions for the limitations and recommendations to e-learning practitioners, management, and leaders.

Freitas and Paredes (2018) conducted a study to assess the background and experience of university faculty members with MOOCs. The faculty expressed their concern on the limited focus of MOOCs on student-centered pedagogies because of the practice in a collection of videos (Daniel et al., 2015). Thus, the first recommendation is that educators and e-learning practitioners should ensure that the introduced MOOCs at their facilities promote personalized and adaptive learning. Course designers, tutors, managers, policymakers, and participants should use existing data on MOOC to improve delivery, educational activities, and learning experiences. Second, the faculty should adopt teaching practices that are concurrent with the introduced courses (Freitas & Paredes, 2018).

Assessment is an essential learning process (Timmis et al., 2015). In their stud,y Costello et al. (2018) found that flawed Multiple Choice Questions (MCQs) adversely affect the reliability and validity of the assessments that lead to inconsistent outcomes for the students. Hence, the third recommendation is that there is a need to eliminate the accreditation, assessment, and certification barriers that hinder the adoption of the new pedagogy. Promoting the credibility of forms of innovative learning and education practices is likely to decrease the students' dropout rate from programs (Daniel et al., 2015).

In the new era, e-learning practitioners need to assess every practice and assumption, apply data when making decisions, consciously collaborate with others, and innovate wherever possible (Mehaffy, 2012). The leadership at institutions of higher learning affects innovation to a significant extent (McGrath et al., 2016). Thus, the leaders should ensure that in universities, there is adequate organization flexibility, quality assurance, and evaluation techniques, institutional support, and financial incentive to support the adoption of innovative learning and education practices. In reference to the ethical challenges associated with the prevalent use of innovative educational and learning, the management and e-learning practitioners should ensure that there exists a desirable balance between the digital and social facets of the practices (Balkin & Sonnevend, 2016). Universities experiencing difficulties in promoting pedagogical change can apply a strategy that has been effective at the London School of Economics (LSE). The strategy includes identifying the environmental factors that cause resistance and limit the sustainability of technological pedagogical change (Bryant, 2018).

FUTURE RESEARCH DIRECTIONS

The increased technological advancements, limited resources, and change in educational paradigms have supported the need to adopt new models of teaching and learning (Pisoni, 2019). A significant number of empirical studies on innovative learning and teaching practices in higher education have been conducted (Brennan et al., 2014; Bryant, 2018; Kalyani & Rajasekaran, 2018; Kovarova & Simkova, 2014; Pilli & Admiraal, 2016; & Stohr & Adawi, 2018). The author of this chapter does not claim to have exhausted the concept of innovative learning and education. Thus, there is still a need for additional research, specifically, systematic reviews on the topic should be conducted. The dynamic nature of the modern world is a significant cause of challenges in the higher education system. This chapter indicates that there is a need for increased research on existing teaching and educational models.

An increase in research on the impact of innovative learning and education practice is likely to promote the adoption of the modern pedagogy at universities in Europe. Hence, the first recommendation for future researchers is to conduct studies assessing the faculty's perception of innovative learning and educational practices. Second, future researchers can conduct studies to assess the transformations in pedagogy that occur as a result of adopting innovative learning and educational practices.

CONCLUSION

The purpose of writing this book chapter was to discuss the innovative learning and education practices that have been implemented in European universities. Traditionally, higher education could only be accessed at specific locations, which made learning geographically limited and spatially fixed. The number of students enrolled at the institutions was determined by the capacity of the infrastructure and human resources availability. The increase in the demand for higher education has created a dilemma for leaders and managers on the manner to support the dissemination of knowledge and accommodating learners at lower costs. The adoption of innovative educational and learning practices has significantly resulted in the transformation of higher education. In Europe, universities have implemented innovative learning and education practices as a technique for responding to modern demand.

The advancement of technologies has made acquiring knowledge to become time and location independent. Universities no longer possess the monopoly of producing and disseminating knowledge. Also, innovative practices have resulted in changes in the instructors' role because the integration of web technologies has increased active learning and access beyond the traditional classrooms. Innovative education (a) supports adaptive techniques that foster personalized content presentation, (b) promotes the dissemination of knowledge at a massive scale, (c) enables reflection and contemplation, and (d) augments traditional pedagogical practices. Innovative education involves the act of leveraging technological advancement that supports the students' ability to achieve program outcomes and increase educators' efficiency. The ideology of innovative education crossover learning, argumentation learning, incidental learning, computational thinking, and understanding through doing science.

The ethical issues associated with innovative education are related to identity, confidentiality, privacy, cheating, and plagiarism. In Europe, innovative educational and learning practices such as MOOCs, blended learning, and flipped classrooms have been adopted in response to the need for effective and flexible educational models in universities. Although the increased adoption of innovative teaching and learning practices has supported the pedagogical change in institutions of higher learning, there is a need for some modifications. The needed modifications are associated with the lack of personalized learning of practices such as MOOCs, accreditation, assessment, and certification barriers, limited accreditation assessment, and ethical issues. There is a need for future researchers to conduct additional studies on innovative and education practices, which support modern pedagogy.

REFERENCES

Adair, D., Alman, S. W., Budzick, D., Grishman, L. M., Mancini, M. E., & Thackaberry, S. (2014). Many shades of MOOCs. *Internet Learning, 3*(1), 53-72. http://citeseerx.ist.psu.edu/viewdoc/download?doi=10.1.1.883.3241&rep=rep1&type=pdf

Alammary, A., Sheard, J., & Carbone, A. (2014). Blended learning in higher education: Three different design approaches. *Australasian Journal of Educational Technology, 30*(4), 440-454. http://doi:10.14742/ajet.693

Aydın, B., & Demirer, V. (2016). Flipping the drawbacks of flipped classroom: Effective tools and recommendations. *Journal of Educational and Instructional Studies in the World, 6*(1), 33–40. https://www.researchgate.net/publication/303813559_FLIPPING_THE_DRAWBACKS_OF_FLIPPED_CLASSROOM_EFFECTIVE_TOOLS_AND_RECCOMENDATIONS

Balkin, J. M., & Sonnevend, J. (2016). The digital transformation of education. In C. Greenhow, J. Sonnevend., & C. Agur (Eds.), *Education and social media: Toward a digital future* (pp. 9-25). MIT Press. https://books.google.com/books?id=6jeNDwAAQBAJ&printsec=frontcover&dq

Barak, M., Watted, A., & Haick, H. (2016). Motivation to learn in massive open online courses: Examining aspects of language and social engagement. *Computers & Education, 94*, 49-60. http://doi:10.1016/j.compedu.2015.11.010

Bennett, S., Bishop, A., Dalgarno, B., Waycott, J., & Kennedy, G. (2012). Implementing web 2.0 technologies in higher education: A collective case study. Computers *& Education, 59*(2), 524–534. http://doi:10.1016/j.compedu.2011.12.022

Bergfjord, O. J., & Heggernes, T. (2016). Evaluation of a "flipped classroom" approach in management education. *Journal of University Teaching & Learning Practice, 13*(5), 1–15. http://ro.uow.edu.au/jutlp/vol13/iss5/17

Bralic, A., & Divjak, B. (2018). Integrating MOOCs in traditionally taught courses: Achieving learning outcomes with blended learning. *International Journal of Educational Technology in Higher Education, 15*(1), 1-16. http://doi:10.118641239-017-0085-7

Brennan, J., Ryan, S., Ranga, M., Broek, S., Durazzzi, N., & Kamphuis, B. (2014, February 2014). *Study on innovation in higher education.* http://eprints.lse.ac.uk/55819/

Bryant, P. (2018). Making education better: Implementing pedagogical change through technology in a modern institution. In A. Zorn, J. Haywood, & J. Glachant (Eds.), *Higher education in the digital age: Moving academia online* (pp. 35-50). Edward Elgar Publishing. https://www.e-elgar.com/shop/higher-education-in-the-digital-age

Burke, P. F., Schuck, S., Aubusson, P., Kearney, M., & Frischknecht, B. (2017). Exploring teacher pedagogy, stages of concern and accessibility as determinants of technology adoption. *Technology, Pedagogy, and Education, 27*(2), 149–163. http://doi:10.1080/1475939x.2017.1387602

Cogill, J. (2010). A model of pedagogical change for the evaluation of interactive whiteboard practice. In M. Thomas & E. C. Schmid (Eds.), *Interactive whiteboards for education: Theory, research, and practice* (pp. 162–178). Information Science Reference. http://doi:10.4018/978-1-61520-715-2.ch011

Conole, G. (2016). MOOCs as disruptive technologies: Strategies for enhancing the learner experience and quality of MOOCs. *Revista de Educación a Distancia, 50*(2), 1-7. http://doi:10.6018/red/50/2

Costello, E., Holland, J., & Kirwan, C. (2018). The future of online testing and assessment: Question quality in MOOCs. *International Journal of Educational Technology in Higher Education, 15*(1), 5-10. http://doi:10.118641239-018-0124-z

Daniel, S. J., Vazquez Cano, E., & Gisbert, M. (2015). The future of MOOCs: Adaptive learning or business model? *Universities and Knowledge Society Journal, 12*(1), 64-73. http://doi:10.7238/rusc.v12i1.2475

European Commission. (2010). *Communication from the commission. Europe 2020: A strategy for smart, sustainable, and inclusive growth.* http://ec.europa.eu/eu2020/pdf/COMPLET%20EN%20BARROSO%20%20%20007%20-%20Europe%20 2020%20-%20EN%20version.pdf

Ferguson, R., & Sharples, M. (2014). Innovative pedagogy at massive scale: Teaching and learning in MOOCs. *Lecture Notes in Computer Science, 2014*(2014), 98–111. http://doi:10.1007/978-3-319-11200-8_8

Freitas, A., & Paredes, J. (2018). Understanding the faculty perspectives influencing their innovative practices in MOOCs/SPOCs: A case study. *International Journal of Educational Technology in Higher Education, 15*(1), 5-10. http://doi:10.118641239-017-0086-6

Garcia, E. B. J., Tenorio, S. G. C., & Ramirez, M. S. (2015). Self-motivation challenges for student involvement in the open educational movement with MOOC. *Universities and Knowledge Society Journal, 12*(1), 91. http://doi:10.7238/rusc.v12i1.2185

George-Walker, L. D., & Keeffe, M. (2010). Self-determined blended learning: A case study of blended learning design. *Higher Education Research & Development, 29*(1), 1–13. http://doi:10.1080/07294360903277380

Hall, J., Chamblee, G., & Slough, S. (2013). An examination of interactive whiteboard perceptions using the concerns-based adoption model stages of concern and the apple classrooms of tomorrow model of instructional evolution. *Journal of Technology and Teacher Education, 21*, 301–320. https://eric.ed.gov/?id=EJ1016990

Heijstra, T. M., & Siguroardottir, M. S. (2017). The flipped classroom: Does viewing the recordings matter? *Active Learning in Higher Education, 2017*(2017), 1-13. http://doi:10.1177/1469787417723217

Henderson, M., Selwyn, N., Finger, G., & Aston, R. (2015). Students' everyday engagement with digital technology in university: Exploring patterns of use and 'usefulness'. *Journal of Higher Education Policy and Management, 37*(3), 308-319. http://doi:10.1080/1360080X.2015.1034424

Kalyani, D., & Rajasekaran, K. (2018). Innovative teaching and learning. *Journal of Applied and Advanced Research, 3*(1), 23-25. http://doi:10.21839/jaar.2018.v3S1.162

Kanwar, A., & Uvalic-Trumbic, S. (2015). *A basic guide to open educational resources. Vancouver: Commonwealth learning.* http://oasis.col.org/bitstream/handle/11599/36/2015_UNESCO_COL_A-Basic-Guide-to-OER.pdf?sequence=6&isAllowed=y

Kintu, M. J., Zhu, C., & Kagambe, E. (2017). Blended learning effectiveness: The relationship between student characteristics, design features, and outcomes. *International Journal of Educational Technology in Higher Education, 14*(1), 7-14. http://doi:10.118641239-017-0043-4

Kirkwood, A., & Price, L. (2013). Technology-enhanced learning and teaching in higher education: what is "enhanced" and how do we know? A critical literature review. *Learning, Media and Technology, 39*(1), 6–36. http://doi:10.1080/17439884.2013.770404

Kop, R., Fournier, H., & Mak, J. (2011). A pedagogy of abundance or a pedagogy to support human beings? Participant support on massive open online courses. *International Review of Research in Open and Distance Learning, 12*(7), 74–93. doi:10.19173/irrodl.v12i7.1041

Kovarova, P., & Simkova, G. (2014). Evidence-based learning approach in evaluation of information literacy education. In *European Conference on Information Literacy* (vol. 492, pp. 560-569). Springer. http://doi:10.1007/978-3-319-14136-7_59

Lasakova, A., Bajzikova, L., & Dedze, I. (2017). Barriers and drivers of innovation in higher education: Case study-based evidence across ten European universities. *International Journal of Educational Development, 55*, 69–79. http://doi:10.1016/j.ijedudev.2017.06.002

Lowenthal, P., & Hodges, C. (2015). In search of quality: Using quality matters to analyze the quality of massive, open, online courses. *The International Review of Research in Open and Distributed Learning, 16*(5), 83–101. doi:10.19173/irrodl.v16i5.2348

Margaryan, A., Bianco, M., & Littlejohn, A. (2015). Instructional quality of massive open online courses. *Computers and Education, 80*, 77–83. http://doi:10.1016/j.compedu.2014.08.005

McGrath, C. H., Hofman, J., Bajzikova, L., Harte, E., Lasakova, A., Pankowska, P., Sosso, S., Belanger, J., & Krivograd, J. (2016, June 31). *Governance and adaptation to innovative modes of higher education provision*. RAND Corporation. https://www.rand.org/pubs/research_reports/RR1571.html

Medina, L. C. (2018). Blended learning: Deficits and prospects in higher education. *Australasian Journal of Educational Technology, 34*(1), 42-56. http://doi:10.14742/ajet.3100

Mehaffy, G. L. (2012). Challenge and change. *EDUCAUSE Review, 47*(5), 25–26. https://er.educause.edu/articles/2012/9/challenge-and-change

Mortensen, C. J., & Nicholson, A. M. (2015). The flipped classroom stimulates greater learning and is a modern 21st-century approach to teaching today's undergraduates. *Journal of Animal Science, 93*(7), 3722–3731. http://doi:10.2527/jas.2015-9087

Murugesan, V. (2019). Innovations in teaching. *Journal of Applied Science and Computations, 6*(1), 2588–2596. https://www.researchgate.net/publication/331149966_Innovations_in_Teaching_Methods

Mykhailyshyn, H., Kondur, O., & Serman, L. (2018). Innovation of education and educational innovations in conditions of modern higher education institution. *Journal of Vasyl Stefanyk Precarpathian National University, 5*(1), 9-16. http://doi:10.15330/jpnu.5.1.9-16

Nat, M. (2015). A flipped classroom model for developing universities in developing countries. In *Global Learn* (pp. 597–604). Association for the Advancement of Computing in Education. https://www.researchgate.net/publication/275638404_A_Flipped_Classroom_Model_for_Developing_Universities_in_Developing_Countries

Norgard, R. T., Burkle, M., Tait, A., Guri-Rosenblit, S., & Canals, L. (2018). Universities of the future: Several perspectives on the future of higher education. In pushing the boundaries of higher education. *International Journal of Educational Technology in Higher Education, 15*(46), 1–10. https://educationaltechnologyjournal.springeropen.com/universitiesofthefuture

O'Flaherty, J., & Phillips, C. (2015). The use of flipped classrooms in higher education: A scoping review. *The Internet and Higher Education, 25*, 85–95. http://doi:10.1016/j.iheduc.2015.02.002

Olivan, B., Masluk, B., Gascon, S., Fueyo Díaz, R., Aguilar-Latorre, A., Artola Magallon, I., & Magallón Botaya, R. (2019). The use of flipped classroom as an active learning approach improves academic performance in social work: A randomized trial in a university. *PloS One, 14*(4), 1-10. http://doi:10.1371/journal.pone.0214623

Pearce, N., Weller, M., Scanlon, E., & Kinsley, S. (2010). Digital scholarship considered: How new technologies could transform academic work. *In Education, 16*(1), 33-44. https://ineducation.ca/ineducation/article/view/44/508

Peterson, A. (2018). *Understanding innovative pedagogies: Key themes to analyze new approaches to teaching and learning*. http://www.oecd.org/officialdocuments/publicdisplaydocumentpdf/?cote=EDU/WKP(2018)8&docLanguage=En

Pilli, O., & Admiraal, W. (2016). A taxonomy of massive open online courses. *Contemporary Educational Technology, 7*(3), 223-240. https://www.researchgate.net/publication/305602116_A_Taxonomy_of_Massive_Open_Online_Courses

Pisanu, F. (2014). Educational innovation and technology: A need for integration. *Perspectives of Innovations, Economics and Business, 14*(2), 103-108. http://doi:10.15208/pieb.2014.12

Pisoni, G. (2019). Strategies for pan-European implementation of blended learning for innovation and entrepreneurship education. *Education Sciences, 9*(2), 124-137. http://doi:10.3390/educsci9020124

Roehling, P. V., Luna, L. M., Richie, F. J., & Shaughnessy, J. J. (2017). The benefits, drawbacks, and challenges of using the flipped classroom in an introduction to psychology course. *Teaching of Psychology, 44*(3), 183–192. http://doi:10.1177/0098628317711282

Schuck, S., & Kearney, M. (2008). Classroom-based use of two educational technologies: A sociocultural perspective. *Contemporary Issues in Technology and Teacher Education: Current Practices, 8*, 394–406. https://citejournal.org/volume-8/issue-4-08/current-practice/classroom-based-use-of-two-educational-technologies-a-sociocultural-perspective/

Simonson, S. R. (2017). To flip or not to flip: What are the questions? *Education Sciences, 7*(3), 71-75. http://doi:10.3390/educsci7030071

Stohr, C., & Adawi, T. (2018). Flipped classroom research: From "black box" to "white box" evaluation. *Educational Sciences, 8*(1), 22-25. http://doi:10.3390/educsci8010022

Szadziewska, A., & Kujawski, J. (2017). Advantages and disadvantages of the blended-learning method used in the educational process at the faculty of management at the university of Gdansk, in the opinion of undergraduate students. In *10th annual international conference of education, research and innovation* (vol. 2017, pp. 3938-3946). http://doi:10.21125/iceri.2017.1051

Timmis, S., Broadfoot, P., Sutherland, R., & Oldfield, A. (2015). Rethinking assessment in a digital age: Opportunities, challenges, and risks. *British Educational Research Journal, 42*(3), 454–476. http://doi:10.1002/berj.3215

Uskov, V., Bakken, J. P., Aluri, L., Rachakonda, R., Rayala, N., & Uskova, M. (2018). Smart pedagogy: Innovative teaching and learning strategies in engineering education. *World Engineering Education Conference, 2018*(2018), 1-6. http://doi:10.1109/edunine.2018.8450962

Wang, K., & Zhu, C. (2019). MOOC-based flipped learning in higher education: Students' participation, experience, and learning performance. *International Journal of Educational Technology in Higher Education, 16*(1), 33-40. http://doi:10.118641239-019-0163-0

Yadav, A., & Patwardhan, A. (2016). Use and impact of web 2.0 tools in higher education: A literature review. In S. Parmar & A. K. Siwach (Eds.), *Academic libraries in electronic environment* (pp. 218-246). Intellectual Foundation. http://doi:10.13140/RG.2.1.2748.6965/1

ADDITIONAL READING

Aljawarneh, S. A. (2020). Reviewing and exploring innovative ubiquitous learning tools in higher education. *Journal of Computing in Higher Education, 32*(1), 57-73. http://doi:10.100712528-019-09207-0

Anshari, M., Alas, Y., & Guan, L. S. (2016a). Developing online learning resources: Big data, social networks, and cloud computing to support pervasive knowledge. *Education and Information Technologies, 21*(6), 1663–1677. doi:10.100710639-015-9407-3

Anshari, M., Alas, Y., Sabtu, N. I. P. H., & Hamid, M. H. S. A. (2016b). Online learning: Trends, issues, and challenges in the big data era. *Journal of e-Learning and Knowledge Society, 12*(1), 121-134. http://doi:10.20368/1971-8829/1003

Anshari, M., Almunawar, M. N., Shahrill, M., Wicaksono, D. K., & Huda, M. (2017). Smartphones usage in the classrooms: Learning aid or interference? *Education and Information Technologies, 22*(6), 3063–3079. http://doi:10.100710639-017-9572-7

Boettcher, J. V., & Conrad, R. M. (2016). *The online teaching survival guide: Simple and practical pedagogical tips*. John Wiley & Sons. https://books.google.com/books?id=QGPSDAAAQBAJ&printsec=frontcover&dq

Brindley, J. (2014). Learner support in online distance education: Essential and evolving. In O. Zawacki-Richter, & T. Anderson (Eds.), *Online distance education: Towards a research agenda* (pp. 287-304). AU Press. https://books.google.com/books?id=9dH9AwAAQBAJ&printsec=frontcover&dq

Carrera, J., & Ramirez-Hernandez, D. (2018). Innovative education in MOOC for sustainability: Learnings and motivations. *Sustainability, 10*(9), 2990-3000. http://doi:10.3390u10092990

Chaurasia, S. S., & Rosin, A. F. (2017). From big data to big impact: Analytics for teaching and learning in higher education. *Industrial and Commercial Training, 49*(7/8), 321–328. http://doi:10.1108/ict-10-2016-0069

Europa Publications. (2019). *The Europa world of learning*. Routledge. https://books.google.com/books?id=plwltAEACAAJ&dq

Greenhow, C., Sonnevend, J., & Agur, C. (2016). *Education and social media: Toward a digital future*. MIT Press. https://books.google.com/books?id=4PwNDAAAQBAJ&printsec=frontcover&dq

Jemni, M., Kinshuk, K., & Khribi, M. K. (Eds.) (2017). *Open education: From OERs to MOOCs*. Springer. https://books.google.com/books?id=MWTUDAAAQBAJ&printsec=frontcover&dq

Nouri, J. (2016). The flipped classroom: For active, effective, and increased learning–especially for low achievers. *International Journal of Educational Technology in Higher Education, 13*(1), 33-44. http://doi:10.118641239-016-0032-z

Varaljai, M. (2016). Establish innovative learning environment by virtual lab concept: An exploratory research in higher education. In *2016 7th IEEE International Conference on Cognitive Infocommunications (CogInfoCom)* (vol. 2016, pp. 000373-000378). IEEE. http://doi:10.1109/CogInfoCom.2016.7804578

Zorn, A., Haywood, J., & Glachant, J. (2018). *Higher education in the digital age: Moving academia online*. Edward Elgar Publishing. https://books.google.com/books?id=vep6DwAAQBAJ&printsec=frontcover&dq

KEY TERMS AND DEFINITIONS

Blended Learning: The process of merging traditional face-to-face with online techniques as an approach for supporting learning.

E-Learning: A comprehensive word that is used to refer to the technology applied in the field of education.

Flipped Classrooms: An innovative practice in the field of education that supports online and in-class learning.

Innovation: The process of formulating or adopting ideas to improve the manner something is done and achieve improved outcomes.

Innovative Education: The act of disseminating knowledge using technology.

Innovative Learning: The process of gaining knowledge through the integration of technology to support and promote the students' experience.

Massive Open Online Course: Online subjects that are available on the web and has open access.

Pedagogy: The action of promoting learning and teaching in an institution.

Web 2.0: A concept that refers to the second generation of internet technologies. It is associated with content creation and an increase in social media activities.

Section 2

Engineering Management Transition and Careers

Chapter 5
Emotional Intelligence in Engineering Management Education:
The Missing Priority

Marilena Antoniadou
https://orcid.org/0000-0002-4025-035X
Manchester Metropolitan University, UK

Mark Crowder
https://orcid.org/0000-0001-8510-261X
Manchester Metropolitan University, UK

George Andreakos
Electricity North West, UK

EXECUTIVE SUMMARY

This chapter aims to increase understanding of how engineering students can benefit from integrating emotional intelligence (EI) into engineering curricula at universities. In particular, it explores the nature of EI and argues for the greater use of EI within engineering education, but also considers the challenges of placing an emphasis on EI within this field. The chapter makes recommendations for how EI skills can be incorporated into engineering education and how universities can seize the opportunity to shape the modern engineer and advance the standing of engineering in the future. The chapter's contribution lies in raising awareness not just about the benefits of integrating EI within engineering education, but also on the challenges that an empathetic behaviour entail. The authors argue that university education needs to be able to prepare graduates with engineering fundamentals and also for success and actual on-the-job EI skills.

DOI: 10.4018/978-1-7998-4063-3.ch005

INTRODUCTION

Emotional intelligence (EI) has been hailed as being an essential skill that individuals need to possess for effective organisational performance. However, this has not been applicable to all industries. Male-dominated professions, such as engineering, have been characterised as being influenced by aggressive management styles, fierce competition, tight profit margins and adversarial relationships (Smithers and Walker, 2000; Walker, 2001). Indeed, engineering workplaces continue to be uneasy environments for professional women, despite the success of women in mathematics and science at university (Gill et al., 2008).

In this masculine environment, emotionality simply implies 'emotional strength', connoting the suppression of, and control over, emotions (Domagalski, 1999). In such professions, men are often deterred by the term 'emotion' itself, as their performance tends to be assessed against certain stereotypes which support and foster decisiveness, toughness, self-reliance, resolution and control (Loosemore and Galea, 2008). On the other hand, engineers are increasingly challenged to emotionally engage with a broad range of internal and external stakeholders. Indeed, recent research (Boyatzis, Rochford and Cavanagh, 2017; Walther, Miller and Sochacka, 2017) positions EI as critical to engineers. However, current efforts to educate emotionally intelligent engineers are hindered by the lack of a conceptually cohesive understanding of, and language for, applying EI to engineering education.

Therefore, whilst research supports the need to nurture functioning interpersonal relationships in engineering (and, by extension, male-dominant professions), many efforts to this effect are hampered by masculinity stereotypes permeating the industry. Within this enduring, albeit changing, sector engineering offers a unique context to demonstrate the 'trainability' of EI to engineering students, an area that this chapter aims to illustrate.

EMOTIONAL INTELLIGENCE (EI)

EI suggests that some people might be more socially and emotionally effective than others in certain aspects of life (Goleman, 1995; Salovey and Mayer, 1990). Goleman, at the beginning of his book on EI, cites Aristotle's words: '[A]nyone can become angry – that is easy. But to be angry with the right person, to the right degree, at the right time, for the right purpose and in the right way – this is not easy' (Goleman, 1995: ix). For Aristotle, those people who know their emotions and know how to deal with them at the right time have a significant advantage in all aspects of their lives, which manifests as the first reference of EI and the importance of emotion in human relationships (Langley, 2000). Similarly, EI involves an array of capabilities, competencies and skills that Goleman categorised in five domains: self-awareness, self-regulation and persistence, motivation, empathy and social skills to handle relationships (Goleman, 1995). Essentially, the first three are personal competencies that determine how people manage themselves, while the remaining two are social competencies which determine how they manage relationships. In particular, empathy – the ability to know and experience the emotions of another person (Duan and Hill, 1996) – is already recognised by the engineering education literature as a core skill, practice orientation and way of being (Walther et al., 2017) and involves showing that the other person's thoughts and emotions are understood and communicated. As such, engineers face challenges when they are to manage project groups and be leaders for organisations because such positions demand skills in social competence and empathy (Rasoal, Danielsson and Jungert, 2012).

In this respect, EI has been proposed as an intrinsically important element in engineering education so that graduates are not only prepared with engineering fundamentals, but also for success and actual on-the-job skills (Riemer, 2003). More specifically, construction firms are seeking graduates with exceptional management and leadership skills to be able to improve interpersonal relationships and to successfully fulfil projects (Dulaimi, 2005), highlighting the influence of academic education and formal training on engineers' behaviour. Moreover, the skills that engineering graduates ought to have include interpersonal skills, teamwork, and motivation (Dulaimi and Langford, 1999). It should be noted that EI should not be considered as a replacement for knowledge, ability or job skills, but instead it enhances work skills. However, Goleman has highlighted that EI abilities were about four times more important than an individual's IQ in determining professional success and prestige – even for those with a scientific background (Goleman, 1998).

EI and Engineering Education

The engineering profession has a long and distinguished history. The word engineer originates from the word *ingeniator*, meaning ingenious in Latin. Engineers have historically been the primary designer and maker of tools (early engineering achievements go as far back as 5500 BC with the use of metal tools and weapons), their role has evolved with the advance of technology from an independent, self-sufficient and highly motivated inventor to an inter-dependent team member of the business world (Yurtseven, 2002). The perception of the engineer has also been transformed in the eyes of society from a know-it-all inventor and tinkerer to a highly skilled and narrowly specialised technical expert.

So, why is Engineering, in particular, the subject of this chapter (apart from the clear link with the book's theme)? Firstly, recent evidence by the Harvard Business Review, shows that thirty-four out of the 100 best performing Chief Executive Officers in the world, in terms of market capitalization of their companies, have undergraduate or graduate degrees in engineering (HBR, 2018). Part of the explanation is that engineering breads an ethos of building things that work and it teaches people to try to do things efficiently and eloquently with reliable outcomes. Secondly, the second most popular course for male students who choose to study in the UK (second most popular destination for higher education studies in the world) is engineering and technology (Bhardwa, 2018). Yet, for females this is not a popular degree, as the engineer's stereotypical negative image of the socially inept genius can inhibit student recruitment (Yurtseven, 2002). Moreover, there is little evidence that EI is part of the engineering curriculum, in any country.

A growing stock of studies suggest that EI is important for functioning interpersonal relationships (Caruso and Salovey, 2004; Lopes et al., 2005) and especially linking EI competencies with an engineer's effectiveness or engagement. Similarly, a series of chapters in the construction literature draw attention to the alleged relevance of EI in this industry (Mo, Dainty and Price, 2006; Thurnell, 2004). Recently, a study by Boyatzis et el., (2017) drew on 30 years of longitudinal studies showing that EI and quality of relationships can be significantly improved with the appropriate pedagogy emphasising the building of one's vision, developmental approaches to EI, developing a shared vision with others, and inspirational coaching. Therefore, if EI has such an impact on outcomes, the interest in enhancing it by means of training or recruiting high EI individuals accelerates. Not only is the business case for EI compelling, it is becoming more and more apparent that because of the complex and dynamic nature of relationships in engineering, good interactions among project actors are crucial to the success of the project. EI, in turn, is thought to be a crucial ability to this end (Mo, Dainty and Price, 2006) and, unsurprisingly, scholars in the realm of construction have turned their interest to EI (Butler and Chinowsky, 2006). In this respect, EI has been proposed as an intrinsically important element in engineering education (Riemer, 2003). From an educational management perspective, it seems plausible to incorporate the concept into engineering curricula at universities.

Planning EI Skills in Engineering Education

One question that schools of engineering may ask is, "can we do more to increase the probabilities that students will develop into good managers and leaders?" One way to do this is to incorporate management skills components into engineering curricula. Indeed, EI needs to be embedded within an organisational system for it to be effective (Fitzgerald, 2003). Similarly, it should also be integrated within the educational structure of engineering studies for it to be effective and reinforced in the learners' minds. Introducing and exposing engineering students to EI literature and skills from an early stage of their degree has proved to be beneficial in facilitating student learning at one particular polytechnic in America (Goleman, 1998). An engineering student will never be able to improve their personal, interpersonal and social skills if they only stick to aspects of the project with which they feel comfortable. Exercises that promote EI skills can be integrated into the course, examples of which are described below.

The major goal of the management skills course should be to teach engineering students the requisite managerial skills to develop their emotional intelligence. 'Skills' imply that one can develop and improve upon them. One approach that has been shown to be effective in helping students to develop and improve their management skills is based on social learning theory (Bandura, 1977), which is based on the premise that people learn by observing and interacting with others, and through observing and copying a 'role model' – known as behaviour modelling (Goldstein and Sorcher, 1974). Social learning theories are considered 'humanist' because control for learning is placed in the hands of the individual and behaviour is shaped by social relationships that influence the individual's beliefs, values and perceptions (Harrison, 2009). This theory proposes that an integration of cognitive and environmental determinants produce a more adequate explanation of human behaviour than focusing on strictly environmental determinants (Manz and Sims, 1980). According to Whetten and Cameron (2002), one useful way of incorporating social learning theory concepts into a management skills training program comprises five components: skill assessment, skill learning, skill analysis, skill practice and skill application. For example, as part of their degree, engineering students may spend a few weeks in the field with experienced manager engineers to learn about the job and what is expected. After the field training, the students can attend workshops to develop the skills they observed in working with their manager engineers models. Based on the pre and post emotional intelligence assessment, Latif (2004) found that the management skills course amongst some pharmacy students resulted in a significant improvement in students' emotional intelligence. A critical highlight of the implementation of this course was the realisation that developing management skills related to emotional intelligence requires a great deal of individual determination and persistence.

The concept of emotional intelligence has also links with reflexivity and reflective abilities (D'Cruz, Gillingham and Melendez, 2007). Asking students to provide insight into their personal emotional reactions, their potential impact on the self and others, as well as how they might affect decision-making processes and personal wellbeing are key elements of reflective practice (Wilson, Ruch, Lymberry and Cooper, 2008). Incorporating a reflective practice has the capacity to help students make connections between the course content and professional practice, to develop their critical thinking abilities and facilitate insights into personal motivations, assumptions and beliefs (Knott and Scragg, 2008). The capacity to manage emotional reactions effectively, frequently in complex care settings, is central to the role of a manager. Therefore, supplementing communication skills across the curricula, and not in a standalone module, has more chances of contributing to students' higher EI levels. More particularly, this can include the delivery of oral presentations in engineering studies and incorporating communication and presentation skills in the marking structure of reports so that students treat them more seriously.

Experiential learning approaches, which involve embedding the student in the actual experience of communication, with opportunities for debriefing and re-application, provide opportunities for the development of self-awareness. This theory is most closely associated with the work of Kolb (1984), which was in turn based upon the earlier works of Dewey (1936). For this type of learning to apply, the learner must firstly engage in a concrete experience, what then follows is reflection on that experience, the learner will then go through a process of abstract conceptualisation in which they draw conclusions from what has been learnt (i.e. what worked, what did not work), and finally, they apply or test what has been learned – active experimentation. These four stages make up the learning cycle. According to Kolb (1984), one needs to complete all four stages, to learn from experience. The inclusion of any experience or activity into the curriculum is relevant here. Also, studying abroad (Andreasen and & Wu, 1999), learning through real case studies (Kreber, 2001), can further be examples of experiential learning. In the classroom, videotape playbacks of oral presentations also stimulate reflection in the student, to help students build on past learning. In addition, role-play can encourage self-awareness, while role reversal will contribute to the students' understanding of empathy, of knowing how the other side perceives engineers. Indeed, such context-specific role-play will help to strengthen those skills within the engineering framework. Mentoring and on the job training of young, technically proficient engineers can also be an effective experiential approach (Farr and Brazil, 2009). Invite young engineers, as part of the training process, to observe or participate in meetings can help young engineers to develop soft skills, whilst it exposes them to several levels of responsibilities.

Practical applications to enhance EI amongst engineering education can also be enhanced with the use of Problem-Based Learning (PBL) to underpin the curriculum. This is a key pedagogical approach that uses a problem as a trigger for students to develop solutions, whilst learning from the process and using evidence to support their decisions (Klegeris and Hurren, 2011). Typically, PBL begins with an unstructured real-world problem which the students need to define and bring into focus before it can be explored (Aditomo et al., 2013). Students must identify what knowledge they need to solve the problem and must take the lead in seeking out and obtaining this. Hence, the key feature of PBL is that the purpose is the acquisition of new knowledge, rather than the application of existing knowledge (Aditomo et al., 2013). Thus, the pedagogical emphasis is firmly upon the process that students go through when solving the problem – the process is more important than the results (Klegeris and Hurren, 2011). It is therefore a fundamentally different approach to education than the traditional 'chalk and talk' approach.

The benefits of PBL, such as the development of social and cognitive skills (Leal Filho et al, 2016), have been recognised in engineering education, where it has a long and well-established pedigree (Edström and Kolmos, 2014). However, despite this, there are contradictions in the literature. For instance, Buchanan et al. (2016) argue that PBL is effective at developing students' critical thinking and problem-solving skills, whereas Koh et al. (2008) disagree. Moreover, whilst challenging the benefits of PBL, Kirschner et al. (2006) argue that its effectiveness is outweighed by the fact that it can overload students. Nevertheless, PBL encourages students to learn and is highly motivational because students can see the link between the problems they are trying to solve and real-world applications (Barrows, 1996). Moreover, the nature of problem-based studies means that there may be more than one solution (Cotič and Zuljan, 2009). This, of course, is ideal from the context of EI.

Key to the success of the approach is the need for tutors to be flexible and to respond quickly to differences in the ability and personality of their students (Savin-Baden, 2007). It is also crucial for assessments to be carefully designed so they are appropriate and align with the learning processes facilitated by the tutors (Hendry et al., 1999). Moreover, success depends on the quality of the problems that are given to students (O'Grady, 2013). Perhaps the most important success factor is the learning environment itself. This must be appropriate to the problems under investigation, teaching methods must support reflection and collaboration, and there must be adequate independent study time built into the programme (Hendry et al., 1999).

CHALLENGES OF INTEGRATING EI IN THE ENGINEERING CONTEXT

Despite the many benefits that EI can provide to engineering management, there are still the contextual circumstances of the profession that make it so unique and impose challenges to the implementation of the concept in engineering education. As already mentioned, the construction and engineering industry is one of the most male-dominated and aggressive industries (Smithers and Walker, 2000). Nevertheless, the impact of expressing and displaying anger, a negatively-perceived emotion, has yielded positive organisational outcomes and perceived effectiveness. For example, if anger were an inappropriate emotion to display in this particular profession, it would be questionable whether project managers would enact it to be effective in their job. Smyth vividly described construction as a profession that 'is based upon the street fighting man', where 'the pressure is great, (and) verbal abuse and shouting are the weapons to instil fear and maintain power in the office corridor' (Smyth, 2000: 12–13). This delineates the parameters within which the typical management style in construction tends to fall. To a significant degree, Lindebaum and Fielden (2010) argue that it is actually good to display anger as such behaviours may be a prerequisite for surviving in this sector, especially against the backdrop that engineering and construction are characterised by aggressive/authoritative management styles, adversarial relation-ships, tight profit margins, fierce competition and the imperative to be able to respond to extreme short-term pressures at work. Hence, it is unsurprising that this results in the industry's preference to recruit assertive candidates, who are perhaps better able to cope with the 'survival of the fittest' maxim (Cartwright and Gale, 1995). Given these unique features of the profession, the argument that the expression of anger suggests lack of emotional regulation and is persistently associated with leader ineffectiveness, as posited by popular as well more scientifically inclined writers (Goleman, 1998), becomes a rather untenable position. However, the management of this anger (and other negatively-perceived emotions) is a skill heavily embedded in emotionally intelligent people. As Lewis (2000) argues, in specific organisational contexts choosing the appropriate emotions to express reflects a leader's ability to respond in an effective way.

CONCLUSION

This chapter has argued for the need to integrate EI skills in the contemporary engineering education, as an increasing prerequisite to become better communicators, team-workers, and reflective and empathic practitioners of engineering. Enhanced integration of EI skills will, in turn, enhance the work skills of graduates, particularly with regard to communication skills. Ultimately, EI skills can be integrated into university teaching, whether it is at the basic instructional level, social and experiential learning, in a work-based learning structure or as a part of PBL-based education. The incorporation of several components of the fundamentals of EI in engineering curricula will facilitate advanced communication skills in order to allow young engineers to integrate and adopt to the contemporary issues of the modern workplace (e.g. increasing mental health issues, conflict, work-life balance etc.). However, given the traditionalist nature of many engineering curricula, this may take some time before change is evidenced. It is also recognised that to change the climate of this male-dominated profession will not happen over-night. Educators and engineers will need to be shown both by word and example that a judicious blend of hard and soft skills is needed to ensure long term success. The earlier the development process is started, the more time is available to grow into a leadership role. As educators ourselves, we must never forget our obligation to younger students of leadership and provide the mentorship needed for success.

REFERENCES

Aditomo, A., Goodyear, P., Bliuc, A.-M., & Ellis, R. A. (2013). Inquiry-based learning in higher education: Principal forms, educational objectives, and disciplinary variations. *Studies in Higher Education*, *38*(9), 1239–1258. doi:10.1080/03075079.2011.616584

Andreasen, R. J., & Wu, C. H. (1999). Study abroad program as an experiential capstone course: A proposed model. *Journal of International Agricultural and Extension Education*, *6*(2), 69–78. doi:10.5191/jiaee.1999.06209

Bandura, A. (1977). *A Social Learning Theory. Prentice Hall.*

Barrows, H. S., & Tamblyn, R. M. (1980). *Problem-based learning: An approach to medical education*. Springer.

Bhardwa, S. (2018). *The least and most popular undergraduate courses in the UK*. Available online: https://www.timeshighereducation.com/student/news/least-and-most-popular-undergraduate-courses-uk

Boyatzis, R., Rochford, K., & Cavanagh, K. V. (2017). Emotional intelligence competencies in engineer's effectiveness and engagement. *Career Development International*, *22*(1), 70–86. doi:10.1108/CDI-08-2016-0136

Buchanan, S., Harlan, M. A., Bruce, C. S., & Edwards, S. L. (2016). Inquiry based learning models, information literacy, and student engagement: A literature review. *School Libraries Worldwide*, *22*(2), 23–39.

Butler, C. J., & Chinowsky, P. S. (2006). Emotional intelligence and leadership behavior in construction executives. *Journal of Management Engineering*, *22*(3), 119–125. doi:10.1061/(ASCE)0742-597X(2006)22:3(119)

Cartwright, S., & Gale, A. G. (1995). Project management: Different gender, different culture? *Leadership and Organization Development Journal*, *16*(4), 12–16. doi:10.1108/01437739510089058

Caruso, D. R., & Salovey, P. (2004). *The Emotionally Intelligent Manager: How to Develop and Use the Four Key Emotional Skills of Leadership*. Jossey-Bass.

Cotič, M., & Zuljan, M. V. (2009). Problem-based instruction in mathematics and its impact on the cognitive results of the students and on affective-motivational aspects. *Educational Studies*, *35*(3), 297–310. doi:10.1080/03055690802648085

D'Cruz, H., Gillingham, P., & Melendez, S. (2007). Reflexivity, its meanings and relevance for social work: A critical review of the literature. *British Journal of Social Work*, *37*(1), 73–90. doi:10.1093/bjsw/bcl001

Domagalski, T. A. (1999). Emotion in organizations: Main currents. *Human Relations*, *52*(6), 833–852. doi:10.1177/001872679905200607

Duan, C., & Hill, C. E. (1996). The current state of empathy research. *Journal of Counseling Psychology*, *43*(3), 261–274. doi:10.1037/0022-0167.43.3.261

Dulaimi, M. F. (2005). The influence of academic education and formal training on the project manager's behavior. *Journal of Construction Research*, *6*(01), 179–193. doi:10.1142/S1609945105000328

Dulaimi, M. F., & Langford, D. (1999). Job behaviour of construction project managers: Determinants and assessment. *Journal of Construction Engineering and Management*, *125*(4), 256–264. doi:10.1061/(ASCE)0733-9364(1999)125:4(256)

Edström, K., & Kolmos, A. (2014). PBL and CDIO: Complementary models for engineering education development. *European Journal of Engineering Education*, *39*(5), 539–555. doi:10.1080/03043797.2014.895703

Farr, J. V., & Brazil, D. M. (2009). Leadership skills development for engineers. *Engineering Management Journal*, *21*(1), 3–8. doi:10.1080/10429247.2009.1143 1792

Fitzgerald, M. (2003). Corporate EQ: Creating an emotionally-intelligent workplace. *HR Professional*, *20*(1), 46–49.

Gill, J., Sharp, R., Mills, J., & Franzway, S. (2008). I still wanna be an engineer! Women, education and the engineering profession. *European Journal of Engineering Education*, *33*(4), 391–402. doi:10.1080/03043790802253459

Goldstein, A. P., & Sorcher, M. (1974). *Changing supervisor behavior*. Pergamon.

Goleman, D. (1995). *Emotional intelligence*. Bloomsbury.

Goleman, D. (1998). *Working with Emotional Intelligence*. Bloomsbury Publishing.

Harrison, R. (2009). *Learning and development* (5th ed.). Chartered Institute of Personnel and Development.

Harvard Business Review. (2018). The Best Performing CEOs in the World. *Harvard Business Review*. https://hbr.org/2018/11/the-best-performing-ceos-in-the-world-2018

Hendry, G. D., Frommer, M., & Walker, R. A. (1999). Constructivism and Problem-based Learning. *Journal of Further and Higher Education*, *23*(3), 369–371. doi:10.1080/0309877990230306

Kirschner, P. A., Sweller, J., & Clark, R. E. (2006). Why minimal guidance during instruction does not work: An analysis of the failure of constructivist, discovery, problem-based, experiential, and inquiry-based teaching. *Educational Psychologist*, *41*(2), 75–86. doi:10.120715326985ep4102_1

Klegeris, A., & Hurren, H. (2011). Impact of problem-based learning in a large classroom setting: Student perception and problem-solving skills. *Advances in Physiology Education*, *35*(4), 408–415. doi:10.1152/advan.00046.2011 PMID:22139779

Knott, C., & Scragg, T. (2008). *Reflective practice in social work*. Learning Matters.

Koh, G. C.-H., Khoo, H. E., Wong, M. L., & Koh, D. (2008). The effects of problem-based learning during medical school on physician competency: A systematic review. *Canadian Medical Association Journal*, *178*(1), 34–41. doi:10.1503/cmaj.070565 PMID:18166729

Kolb, D. (1984). *Experiential Learning*. Prentice Hall.

Kreber, C. (2001). Learning experientially through case studies? A conceptual analysis. *Teaching in Higher Education*, *6*(2), 217–228. doi:10.1080/13562510120045203

Langley, A. (2000). *New essays concerning human understanding*. Macmillan & Co. Ltd.

Latif, D. A. (2004). Using emotional intelligence in the planning and implementation of a management skills course. *Pharmacy Education*, *4*(2), 81–89. doi:10.1080/15602210410001701685

Leal Filho, W., Shiel, C., & Paco, A. (2016). Implementing and operationalising integrative approaches to sustainability in higher education: The role of project-oriented learning. *Journal of Cleaner Production*, *133*, 126–135. doi:10.1016/j.jclepro.2016.05.079

Lindebaum, D., & Fielden, S. (2011). 'It's good to be angry': Enacting anger in construction project management to achieve perceived leader effectiveness. *Human Relations*, *64*(3), 437–458. doi:10.1177/0018726710381149

Loosemore, M., & Galea, N. (2008). Genderlect and conflict in the Australian construction industry. *Construction Management and Economics*, *26*(2), 125–135. doi:10.1080/01446190701798810

Lopes, P. N., Salovey, P., Côté, S., & Beers, M. (2005). Emotion regulation abilities and the quality of social interaction. *Emotion (Washington, D.C.)*, *5*(1), 113–118. doi:10.1037/1528-3542.5.1.113 PMID:15755224

Manz, C. C., & Sims, H. P. Jr. (1980). Self-management as a substitute for leadership: A social learning theory perspective. *Academy of Management Review*, *5*(3), 361–367. doi:10.5465/amr.1980.4288845

Mo, Y., Dainty, A., & Price, A. (2006). The relevance of EQ to construction project management education and practice: an investigative framework. In D. Boyd (Ed.), *Proceedings of the 22nd Annual ARCOM Conference*, (Vol. 2, pp. 823–831). Association of Researchers in Construction Management.

O'Grady, M., Barrett, G., Barrett, T., Delaney, Y., Hunt, N., Kador, T., & O'Brien, V. (2013). Reflecting on the need for problem triggers in multidisciplinary PBL. *AISHE-J: The All Ireland Journal of Teaching and Learning in Higher Education*, *5*(2), 1–21.

Raiden, A. B., Dainty, A., & Neale, R. H. (2004). Current barriers and possible solutions to effective project team formation and deployment within a large construction organization. *International Journal of Project Management*, *22*(4), 309–316. doi:10.1016/j.ijproman.2003.08.002

Rasoal, C., Danielsson, H., & Jungert, T. (2012). Empathy among students in engineering programmes. *European Journal of Engineering Education*, *37*(5), 427–435. doi:10.1080/03043797.2012.708720

Salovey, P., & Mayer, J. (1990). Emotional intelligence. *Imagination, Cognition and Personality*, *9*(3), 185–211. doi:10.2190/DUGG-P24E-52WK-6CDG

Savin-Baden, M. (2007). *A practical guide to problem-based learning online*. Routledge. doi:10.4324/9780203938140

Smithers, G. L., & Walker, D. H. T. (2000). The effect of the workplace on motivation and demotivation of construction professionals. *Construction Management and Economics*, *18*(7), 833–841. doi:10.1080/014461900433113

Smyth, H. (2000). *Marketing and Selling Construction Services*. Blackwell.

Thurnell, D. (2004). Emotional intelligence and the value management facilitator. In F. Khosrowshahi (Ed.), *Proceedings of the 20th Annual ARCOM Conference*, (Vol. 2, pp. 918–922). Association of Researchers in Construction Management.

Walker, M. (2001). Engineering identities. *British Journal of Sociology of Education*, *22*(1), 75–89. doi:10.1080/01425690020030792

Walther, J., Miller, S. E., & Sochacka, N. W. (2017). A model of empathy in engineering as a core skill, practice orientation, and professional way of being. *Journal of Engineering Education*, *106*(1), 123–148. doi:10.1002/jee.20159

Whetten, D. A., & Cameron, K. S. (2002). *Developing Management Skills* (5th ed.). Prentice Hall.

Wilson, K., Ruch, G., Lymbery, M., & Cooper, A. (2008). *Social work: An introduction to contemporary practice*. Pearson.

Yurtseven, H. O. (2002). How does the image of engineering affect student recruitment and retention? A perspective from the USA. *Global Journal of Engineering Education*, *6*(1), 17–23.

Chapter 6
Engineering to Management Transition:
Challenges and How Education Might Assist

Leonidas Efthymiou
https://orcid.org/0000-0001-8099-5885
University of Nicosia, Cyprus

Epaminondas Epaminonda
University of Nicosia, Cyprus

Despo Ktoridou
University of Nicosia, Cyprus

EXECUTIVE SUMMARY

This chapter identifies the main challenges in the transition from engineering to management and discusses how management education may assist in this transition. Mixed methods were used to achieve the above. Initially, two focus groups were conducted, and at a later stage, 126 engineers reported through a questionnaire the most common challenges in the transition from engineering to management and then a few were also interviewed. Results demonstrate that skills such as delegation, communication, convincing, coaching, and guiding others pose important challenges. In relation to the second inquiry, it is proposed that, other than management theory, offering examples, opportunity to practice with feedback, and case-based learning can help minimise the challenges. Also, leadership skills, such as delegation, developing personality, cross-cultural understanding, and managing diversity, can be strategically used to facilitate learning in the field of engineering and better prepare engineers in their transition to management.

DOI: 10.4018/978-1-7998-4063-3.ch006

INTRODUCTION

The transition to a management position is a common characteristic of an engineering career (Lloyd, 1991). Engineers who demonstrate competence in carrying out their job often take up extra responsibilities and guide and direct others in performing their jobs. Gradually, they are identified as possible managers and, occasionally, they eventually get promoted. Indicatively, according to a study by executive search firm Spencer Stuart, 33% of S&P 500 CEOs held engineering degrees (Lutchen, 2016).

The transition from a position that involves purely engineering tasks to management is not easy however (Roberts & Biddle, n.d.). Several scholars have identified the challenges of this changeover, which are attributed to the kind of tasks managers and engineers have to tackle, as well as their different training and mindset. As (Hsu, 2017) pointed out, 'the emphasis on objectivity and facts during engineering education results in problems in dealing with the more vaguely defined problems of management' to conclude that 'the problems [in management] are people problems, communication problems, all the things [engineers have] generally learned to devalue as engineers'.

Lack of education in management is one factor that contributes to engineers lacking 'soft' skills, which are vital for successful management (Seethamraju & Agrawal, 1999). These gaps are usually identified several years after they attain their formal engineering qualifications, in the middle-to-late career phases (Yeh, 2008). To advance further, engineers express a strong interest for management training, whereas many of them seek post-graduate qualifications through a master's degree in management or business administration (Batley, 1998). These observations, reemphasise the need to proactively include management education in engineering studies so that engineers are equipped with the right skills prior to facing transition obstacles.

Moreover, the transition challenges are also attributed to personal and contextual factors. Studies in the UK, USA, Germany, New Zealand, France, Japan and Australia, reveal that engineers, their expressed interests, personalities, priorities, professional satisfaction, attitude towards career advancement, status within society, occupational identity, local industrial relations, cultural factors and organisational policies pose different challenges on this transition (Buchanan, 1989; Florman, 1994; Glover & Kelly, 1987; Hill, 1993; Martínez-León et al., 2018; Rynes, 1987; Sakakibara & Westney, 1995; Williams, 1988). These observations call for localised studies, aiming at exploring the challenges of advancing engineers in their own national, cultural and workplace context.

While previous literature has provided valuable insights regarding the possible management-related challenges of engineers, as well as the role of education in this changeover, two important questions seem to persist: 'what challenges are encountered in the transition from engineering to management in other national contexts' and 'how can engineering management education prepare future engineering managers for this transition'?

To answer these questions, the authors have conducted two different studies in Cyprus, aiming to capture the voice of both engineering students as well as engineering professionals. In Study 1, 20 engineering management master's students with experience in engineering participated in two focus groups, in which they were asked to identify challenges they encountered in their transition from engineering to management and how management education could assist them in overcoming these challenges. In the second study, 126 engineers were asked through a questionnaire to name which challenges they encountered more frequently in the transition from engineering to management. The questionnaire was circulated via professional engineering associations and personal contacts. To our knowledge, this is the first study that explores and combines the opinions of both students (who are already professional engineers) and practicing engineers.

The chapter is organised as follows: first the literature on the challenges posed by the transition of engineers to management is reviewed. After that, the research design is outlined, together with specifics for each study. Then, the analysis discusses the fieldwork results, prior to summarising the implications stemming from the research, along with the article's contribution.

BACKGROUND

Differences Between What Engineers and Managers Do and Perceive at Work

To explore how engineers become managers, along with the difficulties they face in this transition, it is important to highlight how management emerged out of engineering and how the two became progressively different. At the end of the nineteenth and early twentieth centuries, 'Classical Management' originated by engineering managers in the manufacturing, transportation and other industries. Frederick Winslow Taylor (Taylor, 2005), an American mechanical engineer (also known as the 'father' of Scientific Management), spent a great part of his life striving to achieve efficiency on the shop floor. After a long and successful employment at Midvale Steel Company, he moved to Bethlehem Still Company (in 1889) in Pittsburgh where he conducted a series of experiments on improving labor productivity. Likewise, Henry Fayol (Fayol, 1949), a French mining engineer who worked all the way to the position of Managing Director, applied his own principles to turn an almost bankrupt company to a prosperous and successful entity (Cole

& Kelly, 2016). In his writings in 1916, he defined management and referred to key managerial activities such as forecasting, planning, organising, commanding, coordinating and controlling. A hundred years later, these activities are still central in contemporary organisations.

Following Taylor and Fayol, several key scholars expanded the theory, scope and practice of management even further. Scientific Management became 'Administrative Management' by the addition of measurement, people development and performance bonus systems [e.g. 19, 20]; whereas, an academic (sociologist) namely Max Weber (Roth et al., 1978), theorised an organisation form that he named 'Bureaucracy' – with an emphasis on authority structures. The new additions in the body and practice of management gradually replaced the engineers' logic. From analysing, modifying and standardising workflows in the pursuit of efficiency and productivity, management became strategic in the early 20th century. Then, it adopted a more humanistic-democratic context and shifted away from the depersonalised, mechanistic value systems of Scientific Management and Bureaucracy (Cloke & Goldsmith, 2001). Collaboration, vision and reason gradually replaced the typical standardising model of early practicing engineers. Organisations were recognised as complex systems whereas self-managed teams and down-to-top informal leaders freed managers by traditional supervisory, top-down hierarchies. By that time, engineering and business management evolved as independent disciplines and two separate functions.

An engineering job often includes inventing, designing, analysing, building, and testing machines, systems, structures and materials while considering the limitations imposed by practicality, regulation, safety, and cost (U.S. Department of Labor, 2006); (National Society of Professional Engineers, 2006). A management job, on the other hand, includes setting objectives and planning, organising (work and people), motivating and communicating, measuring performance, developing people (Drucker, 1993) and it involves interpersonal roles (figurehead, leader, liaison), informational roles (monitor, disseminator, spokesman) and decisional roles (entrepreneur, disturbance handler, resource allocator, negotiator) ((Mintzberg, 1971)), among others. (Souder, 1983) summarised the differences in what engineers and managers do in Table 1.

The different tasks engineers and engineering managers perform at the workplace also influence how work is perceived. For example, while examining the values differentiating engineers from engineering managers, (Munson & Posner, n.d.) identified that for engineers, management work may not count as real work. 'Managers were perceived as having to "attend numerous meetings, frequently report to upper management, deal with employee problems...", all of which reduce the time for "working on the real problems."' In other words, the differences between what engineers and managers do, is not simply about their daily tasks. It is also about how they understand and perceive work itself. Given these differences, the next section presents and discusses what engineers themselves have considered as challenges during their advancement to management.

Challenges Faced by Engineers in the Transition to Management

Table 1. What engineers and managers do

What Engineers Do	What Managers Do
Emphasise accuracy and mathematical precision	Rely heavily on intuition, make educated guesses and try to be "about right"
Exercise care in applying sound scientific methods on the basis of reproducible data	Exercise leadership in making decisions under widely varying conditions based on sketchy information
Solve technical problems based on their own individual skills	Solve techno-people problems based on skills in integrating the talents and behaviours of others
Work largely through their own abilities to get things done	Work primarily through others to get things done

Source: Souder (1983)

The challenges of engineers during their transition to management may vary according to context and preferences of each individual. The context includes unique characteristics such as the country, culture, status within society, local industrial relations and organisational policies. The preferences include interests, personalities, priorities, professional satisfaction, attitude towards career advancement and occupational identity of each individual (Hood, 1990); (Badawy, 1995); (Howard, 2003); (Bailyn, 1991); (Dean, 2004); (Martin, 2010); (Immen, 2010); (The Economist, 2004). In other words, observations from the UK, USA, Germany, New Zealand, France, Japan and Australia, underline that challenges are by no means universal.

For instance, (Hood, 1990) found that key challenges included: (1) delegating appropriately; (2) ability to communicate skilfully; (3) directing and guiding others; (4) balancing quality and perfection with time and resource constraints. (Martínez-León et al., 2018), on the other hand, noted that building effective relationships was the most difficult challenge. Other challenges identified included, (1) heavy workload; (2) the realisation that they had to get things done through others and thus were dependant on their subordinates; (3) learning to delegate; and (4) the feeling that they were like 'firefighting'. (Badawy, 1995) added the following challenges: (1) determining priorities; (2) determining optimum degree of help to give to individuals; (3) inconclusive meetings; (4) paperwork; and (5) dealing with human resource shortages.

In another comprehensive study, (Howard, 2003) identified nine challenges, which he ranked in terms of difficulty. The three most difficult were: (1) so much going on: the engineering manager role involves balancing many more responsibilities, tasks and priorities than the engineering role; (2) relationship changes: personal relationships, interaction dynamics have changed; and (3) delegation: the challenge of leaving the hands-on technical behind and learning to work through others. The next three challenges were: (4) increased stress and pressure associated with increased responsibility; (5) developing new skills; and (6) finding the time, the staff and other resources to get it done. The last three challenges were: (7) the new guy in management: change from being a technical expert to being new in management and having a lot to learn; (8) organisational issues: in a new organisational level with its associated issues; and (9) choosing the management career path: the concerns before deciding and questions experienced during or after the transition.

An attempt to summarise challenges and put them in categories is shown in Table 2. As may be observed the first three challenges (delegating, communicating, managing people) relate directly to the management roles the engineers have now acquired. The next three (resource constraints, work overload) are challenges that may be encountered in any job, even though here they relate more to the management role. The last three are more personal challenges.

These challenges have a direct implication in the workplace. Consider for example the attempts of a managing engineer (with limited management knowledge and skills) to control a group of engineers who participate in a Community of Practice (CoP) for the development of a new technology. The challenges are likely to influence the entire process, from conception (thinking together and sharing knowledge) to collective execution. Moreover, the work setting becomes even more complex when one considers that challenges may differ by context and individual characteristics. The challenges, nevertheless, also have implications for education, as discussed in the following sections.

How Education Might Assist in Overcoming Engineers' Challenges in the Transition to Management

As explained earlier, the managerial skills required for the transition to management differ significantly from the technically oriented education and training of engineers (Bailyn, 1991). This identification led to emphasising the role of education in assisting the transition of engineers into management since the 1950s (Shepard, 1958). What is interesting to note, nevertheless, is that the same concerns are raised time and again. For instance, in the 1970s, (Badawy, 1995) asked why so many competent engineers

Table 2. Challenges faced by engineers in the transition to management

Challenge Category	Examples of Challenges Mentioned
Delegating	Delegating appropriately; the realisation that they had to get things done through others and thus were dependent on their subordinates; learning to delegate; determining optimal degree of help to give to individuals; delegation: the challenge of leaving the hands-on technical work behind and learning to work through others
Communicating	Ability to communicate skilfully
Managing people	Directing and guiding others; inconclusive meetings
Resource constraints	Balancing quality and perfection with time and resource constraints; determining priorities; dealing with human resource shortages; finding the time, the staff and other resources to get it done
Work Overload	Heavy workload; determining priorities; so much going on
Dealing with unexpected problems	The feeling that they were like 'firefighting'
Relationships	Relationship changes: personal relationships, interaction dynamics have changed
Learning	The new guy in management: change from being a technical expert to being new in management and having a lot to learn
Stress	Increased stress and pressure associated with increased responsibility

Source: Authors

become incompetent managers. This is attributed, he explained, to 'inadequate preparation for careers in management' (p.40). Rather than over-developing their analytical skills (as model builders), education can also consider their managerial skills (as decision makers), he suggested.

In the late 1990s, (Seethamraju & Agrawal, 1999) reported similar deficiencies in Australia: [Although e]ngineers are the second largest group of professionals in Australia, and constitute a significant proportion of the managerial pool… [they] are generally viewed to be unsuccessful in management roles… [due to] inadequacies in engineering education... attitude of engineers towards management roles [and] management education. The study concluded that the probabilities of a successful transition to managerial roles goes through changes in undergraduate engineering education and the acquisition of postgraduate management qualifications, among others.

Similar studies made attempts to identify what business management subjects are needed for training established engineers several years after the attainment of engineering qualifications. Prominent subjects on the list include people-oriented skills (personal and interpersonal management), general management, human resource management, finance, accounting and project management (Batley, 1998).

However, additional training courses for existing engineers do not seem to prevent transition challenges. Rather, they reflect the temporary nature of reactive solutions.

Another approach suggests that integrating leadership learning earlier (e.g. into the technical curriculum of undergraduate or master engineering students) enables a fuller socio-technical development and accelerate a process that would otherwise take decades of trial and error (Rottmann et al., 2019).

But, although there is a widespread consensus among scholars on the vital role of education in rectifying possible challenges, the way that this can be done is still open to debate and that is an area this chapter can contribute to.

METHODS

To evaluate the possible challenges faced by engineers and the type of education that could better prepare them for a transition to management, three research methods and two studies were used as follows:

Study 1 - Focus Groups

In study 1, two (2) focus groups were organised with 20 postgraduate students (10 participants in each group). All postgraduate students are existing engineering practitioners. The focus groups were conducted in November 2018. The method of focus groups was selected to enable interaction between participants, aiming to illuminate key topics and common challenges. The first focus group lasted 75 minutes and the second 90 minutes. The focus groups were recorded and transcribed. Focus groups focused on identifying challenges that participants encountered in their transition from engineering to management and the type of management education could assist them in overcoming these challenges. The researchers decided not to conduct a third focus group once they were confident that the possibility for new insights was saturated (as suggested by (Lincoln & Guba, 1985)). Saturation became evident when the claims and repetitions of focus-groups' members were repetitive and identical.

Study 2 - Questionnaires

In Study 2, an online questionnaire was designed based on the literature review and the focus groups' findings, including the main challenges expressed by focus-groups participants regarding the transition from engineering to management. The main challenges were listed and respondents were asked to rate the extent to which they encountered them in their transition. Five levels were given as choices in the answers (plus one option for the case when the challenge was not encountered). More specifically, the descriptions of each level and the number assigned to each one were the following: Not encountered (0), Encountered to a minimum extent (1), Encountered to some extent (2), Encountered to considerable extent (3), Encountered to significant extent (4), Encountered to the highest level (5). The questionnaire was disseminated to engineering practitioners via Engineer Professional Associations and personal contacts in February 2020. The questionnaire was available online for two months. It is probably worth noting that the last four weeks of availability coincided with the Coronavirus epidemic period which may have affected the response rate. The number of completed questionnaires was 126.

Respondent Validation – Interviews

After the completion of the questionnaire, telephone interviews were conducted with some of the respondents to discuss their answers. The process was conducted in April 2020. The purpose of the interviews was to reconfirm the researchers' interpretation in regard to certain participants' responses. This method is called 'respondent validation', also known as 'member validation' or 'member checking'. This method is according to (Bryman & Bell, 2007), 'a process whereby a researcher provides the people on whom he or she has conducted research with an account of his or her findings and requests feedback on that account'. The interview discussions focused on explaining why participants encountered certain challenges and how previous education could help them avoid those challenges. A total of eight interviews were conducted. An analysis and some quotes are included in the actual analysis.

DATA ANALYSIS

Challenges in the Transition From Engineering to Management

Focus Group Study

The challenges that were identified during the focus groups were the following:

1. Being dependent on others
2. Not being certain of the outcome even if their performance was good
3. Having to convince people
4. Having to manage diverse individuals
5. Not being certain that a specific behaviour will lead to a desired outcome
6. Dealing with unexpected (usually human related) problems

The first challenge that was mentioned by almost all participants was the fact that in management they were dependent on others whereas in engineering they were mostly depending on themselves. As one participant who worked in programming in the past said, *'In my engineering job I was almost wholly depended on myself. If I did a good job in writing the code, I knew it would work, or if it did not, I could fix it. But in management I had to depend on others to get the job done.'*

A similar view was echoed by an electrical engineer turned manager who said: *'In my engineering job I was in charge of choosing and ordering machines and then I was in charge of making sure they work properly. Very rarely did I have people interfere with my work, other than the discussions I had with my manager about the budget. As a manager I have to oversee a group of 10 people and I depend on them. If they do not turn up [to work] or do not do their checks properly I will be in trouble.'*

Not everyone had the same feeling though. A civil engineer noted that he depended on others in both his jobs since many of his civil engineering projects involved other people.

The fact that they depended on other people related to the second challenge that came out: Not being certain of the outcome even if their performance was good. As participants explained, in their engineering jobs the outcome of their work was mostly depended on themselves. If they worked harder or better the outcome would almost invariably be improved. In management though, their planning or thinking about a problem was not a guarantee of success since other people had to work for it – and sometimes things did not go as planned. One participant mentioned one example of a meeting where he had to explain to members of his group a new way of working. He had everything planned, he said, but the meeting took a very different route compared to what he expected when one participant disagreed about the proposals at the very beginning and the others followed him and it was very difficult to get the group back on track.

This related to the third challenge: having to convince people. As one participant said: *'I used to spend a lot of time before meetings to analyse how to convince others about a new idea, a new project, a rule we had to follow. Even when I felt I had the perfect argument, it often did not work on everyone and I had to readjust my approach.'*

This led to the fourth challenge: Having to manage diverse individuals. As one participant said, *'I noticed that different people wanted different kinds of arguments to be convinced. Some were more rational, others more emotional, other had behaviour that fluctuated from day to day.'* Another issue that was raised here was cultural diversity. One participant who worked frequently with multinational teams commented on the way people from different countries exhibited different behaviours.

The fifth challenge – not being certain that a specific behaviour will lead to a desired outcome – was essentially explained as the contrast between engineering, where a change in an input led to a predicted, often proportional effect on output and management where this was not the case with people.

Finally, participants found it a challenge that they had to deal with unexpected problems, usually related to human resources. As one participant noted, '*It was difficult when problems such as absences or conflicts between employees arose. They could not have been planned beforehand and they delayed the job to be done. In many cases I felt like firefighting.*'

Questionnaire and Interview Studies

Questionnaire Demographics

Eighty five point seven percent (85.7%) of the respondents were male and 14.3% female. Thirty one point seven percent (31.7%) of respondents were between the ages of 25 and 34, 41.3% were ages 35-44, 19% 45-54 and 7.9% 55-64. With regard to length of work experience, 1.6% of respondents had less than one year of work experience, 23.8% between one and five, 20.6% six to 10, 28.6% 11-19 and 25.4% over 20. Regarding the interviews which followed the questionnaires, as a method of respondent validation, six participants were men and two were women. One interviewee was above 55 years old, two 45-54, three 35-44 and two 25-34.

Results

Table 3 provides a summary of the questions that asked participants to rate the challenges in moving from engineering to management. The challenges are ranked from highest to lowest average score. The highest score was 4.05 and the lowest 3.06. The challenges are presented and discussed in a descending average score order.

The top three challenges have to do with managing relationships with others. All of them - 'directing and guiding others', 'convincing them to perform a task' and 'managing relationships with them' - are essential elements of management. This finding is in line with focus group findings which highlighted the challenge in moving from the clearly defined tasks of engineering to the people's skills needed in management. As one participant said in the interview stage, '*I found managing others and getting them to do things without complaints very challenging because in contrast to engineering where an established theory always work in the same way, with people you have to adapt your approach because each one is different.*'

The fourth challenge was 'determining priorities'. This challenge had to do with the fact that individuals in positions of management have many ongoing things to take care of in contrast to an engineering position where the typical task was to work on a project that had to be completed. The next three challenges on the list

- 'increased stress and pressure associated with increased responsibility', 'dealing with unexpected problems' and 'work overload' - seem to be related. The increase in workload, which is often associated with increased responsibility and which often includes unexpected problems, leads to increased stress and pressure. This chain of management position characteristics can be quite challenging, as noted by participants.

The next challenge was resource allocation. Here participants stressed the difficulty of deciding on how to allocate available funds between often equally good alternatives or between departments and people who requested them. As one participant said, '*It is quite puzzling to decide between equally appealing or uncertain business opportunities or between middle managers who request the support of their functions.*'

The ninth challenge was 'delegating'. Here participants explained the difficulty of trusting others to perform a task, a common management challenge. The next three of the four challenges that follow have to do with human resource management. 'Appraising employees' was the one that received that higher score of the three. Participants here noted the difficulty of appraising coworkers both in terms of being generally objective but also due to the personal relationship they had with them. One noted, *'[having to appraise coworkers] is a kind of feeling I never had in my engineering career and it is quite hard.'*

The next challenge appearing a bit lower on the list is the fact that problems are often 'less clearly defined' in management. This relates to the fact, according to participants comments, that in engineering the problems were directly related to the task at hand whereas in management it was not clear which of the factors of the external or internal environment had to be dealt with.

The next two human resource related challenges were 'hiring staff' and 'deciding on what rewards to offer to subordinates'. In relation to the former of these challenges, participants noted the difficulty of choosing between often equally good people and in relation to the latter the difficulty of deciding primarily between fixed and variable rewards.

The final challenge that received the lowest score was 'resource constraints'. Here participants agreed that even though it is an important and difficult issue, they did not consider it a challenge in the transition from engineering to management because it a common issue in engineering as well.

Table 3. Challenges in the transition from engineering to management

	Challenge	0	1	2	3	4	5	Average
1	Managing others (Directing and guiding others)	0.0	0.0	12.1	16.7	25.8	45.5	4.05
2	Convincing others to perform a task	0.0	6.1	9.1	6.1	34.8	43.9	4.02
3	Managing relationships with subordinates	0.0	3.0	9.1	13.6	33.3	40.9	4.00
4	Determining priorities	3.0	0.0	6.1	13.6	40.9	36.4	3.98
5	Increased stress and pressure associated with increased responsibility	3.0	4.5	6.1	6.1	40.9	39.4	3.95
6	Dealing with unexpected problems	3.0	7.6	12.1	3.0	24.2	50.0	3.88
7	Work overload	1.5	3.0	16.7	10.6	31.8	36.4	3.77
8	Resource Allocation (deciding how to allocate financial and human resources)	0.0	3.0	16.7	19.7	28.8	31.8	3.70
9	Delegating (entrusting a task or responsibility to another person)	0.0	3.0	18.2	18.2	31.8	28.8	3.65
10	Appraising performance of others	0.0	7.6	13.6	19.7	25.8	33.3	3.64
11	Less clearly defined problems compared to engineering	1.5	4.5	15.2	18.2	34.8	25.8	3.58
12	Hiring staff	6.1	3.0	13.6	22.7	27.3	27.3	3.44
13	Deciding on what rewards to offer to subordinates	6.1	6.1	19.7	15.2	31.8	21.2	3.24
14	Resource constraints (limited financial resources to perform tasks)	3.0	10.6	27.3	13.6	27.3	18.2	3.06

How Management Education Can Assist in the Transition to Management

In the focus-group meetings, participants were also asked to discuss how management education has contributed to dealing with the challenges encountered in the transition from engineering to management.

In relation to the first challenge identified in the focus groups, being dependent on others, participants pointed out that one activity that has helped was working in groups. This gave them the opportunity to collaborate with others in projects which were similar – but not identical – to leading a team. As one participant noted: *'Working in groups has some similarities to leading a team even though it is not of course the same. In groups you exchange opinions and you need to consider others' views but when you are in charge of a team it is a different kind of responsibility.'*

Regarding the second challenge, not being certain of the outcome even if their performance was good, participants could not identify a specific experience in their education that has helped directly, but they made references to case studies they had discussed in their management studies. They noted that by considering the case studies' scenarios they realised the complexity of business decision making and the not so direct linkage between action and result. More specifically, they commented on the several external factors that influence business decisions and outcomes.

The challenge of having to convince others was related, according to the participants, to a few aspects of management education, namely communication, negotiation and presentation. In all cases, participants noted that they found both the theory and the practical exercises useful.

The fourth challenge, having to manage diverse individuals, related, according to the participants, to discussions on personality and cultural differences. Classes on personality illuminated differences between individuals and cultural dimensions differences between cultural groups. Individual assignments that included for instance taking personality tests and cases on cultural differences were also mentioned as helpful experiences.

Participants could not identify something specific in their management studies that related to the fifth – not being certain that a specific behaviour will lead to a desired outcome – and sixth – dealing with unexpected (usually human related) problems – challenges. However, they noted that a number of discussions related indirectly. For example, one participant said: '*Probably we did not say explicitly that stimulus and response in human behaviour are not proportional and that human actions cannot be predicted but the discussions we had about motivation and the diverse factors that influence human behaviour essentially suggested that. Therefore, I would say that these discussions were highly influential for me in relation to how I view human behaviour*".

Other challenges that were identified in the questionnaire study included: determining priorities; resource allocation; less clearly defined problems compared to engineering; hiring, appraising and offering rewards; increased stress and pressure associated with increased responsibility. For the first four challenges, related case studies could assist. The first three challenges would be more related to general management/business case studies whereas the fourth one would be more related to human resource management. Increased stress and pressure is a common complain of many workers recently. Short courses in time or stress management could be of assistance, but also being more prepared for management responsibilities through all the methods mentioned before could also contribute to decreasing the level of stress and pressure felt due to assuming management duties.

CONCLUDING DISCUSSION

This paper aimed to identify main challenges in the transition from engineering to management and discuss how education can assist in this transition.

Regarding the first aim, literature suggests that there are several challenges, most of them relating to the fact that in management most problems are people related. More specifically, the challenges identified were grouped in nine categories, the first three being related to management, the next three to the job itself and the last three to personal challenges. The analysis of focus groups suggested six challenges which related to the management function and the unpredictability of human behaviour. In the questionnaire study, 14 challenges were rated from considerable to significant with the top ones being related again to managing others (directing and guiding others, convincing others to perform a task and managing relationships with subordinates).

Regarding the extent to which management education helps in dealing with the challenges, literature suggests presenting management theory, offering examples and opportunity to practice with feedback. Also, case-based learning seems to be a good way to highlight the complexities of management and develop related skills. More specific suggestions include developing delegation and communication skills and pointing out the difference between engineering and managing people. The analysis of the focus groups suggests that communication exercises, analysis of personality and culture, case studies and practical exercises are elements of a management program that may assist in dealing with common challenges that are encountered in the transition from engineering to management.

This was also evident in the second study as well as the respondent validation that followed. Based on findings, the importance of case-based learning as part of management courses in engineering programs is highlighted. Our suggestion resembles earlier studies, stressing the importance of case-base learning in management (e.g. (Ktoridou & Eteokleous, 2012); (Ktoridou et al., 2016); (Ktoridou et al., 2018)). Management skills in business schools are developed by both theoretical presentation and analysis of relevant theories and their application in practical situations. Traditionally, a widely applied model for skill development in business schools is (Cameron & Whetten, 1983). According to these authors, skill development of management skills would be good to follow five steps. First, aspiring managers need to assess their current level of skills. Then they need to be taught about the principles of the skill to be developed. Third, they need to be provided with examples of appropriate skill performance. Fourth, they need to practice the skill and receive feedback and, fifth, apply it in real life situations.

Overall, it seems that key challenges in the transition from engineering to management relate to the inherent differences between the training of an engineer and a manager and the central role that people and relationships play in managing people. Therefore, in training engineers to become managers a key element of their education should relate to this difference. As Tsu (2017) pointed out, '*Going from being an engineer to being an engineering manager isn't merely a change in title. It's an identity shift*'.

REFERENCES

Badawy, M. K. (1995). *Developing Managerial Skills in Engineers and Scientists* (2nd ed.). John Wiley & Sons.

Bailyn, L. (1991). The hybrid career: An exploratory study of career routes in R&D. *J. Eng. Tech. Manage.*, *8*, 1–14.

Batley. (1998). Management training of professional engineers in New Zealand. *Journal of Euro-pean Industrial Training, 22*(7), 309–312.

Bryman, A., & Bell, E. (2007). *Business Research Methods* (revised edition). Oxford University Press Inc.

Buchanan, R. A. (1989). *The Engineers - A History of the Engineering Profession in Britain I750 - 19145*. Jessica Kingsley publishers.

Cameron, K. S., & Whetten, D. A. (1983). A model for teaching management skills. *The Oranizational Behavior Teaching Journal*, *8*(2), 21–27.

Cloke, K., & Goldsmith, J. (2001). *The End of Management and the Rise of Organisational Democracy.* Wiley.

Cole, G. A., & Kelly, P. (2016). *Management Theory and Practice* (Custom Edition). Cengage.

Dean, E. (2004). Can management be taught? If so, what should management education curricula in-clude and how should the process be approached? *Management Decision*, *42*(3/4), 439–453. doi:10.1108/00251740410523240

Drucker, P. F. (1993). *Management: Tasks, Responsibilities, Practices*. HarperBusiness.

Fayol, H. (1949). General and Industrial Management. Pitman. (Original publication 1916)

Florman, S. C. (1994). *The Introspective Engineer*. St. Martin's Press.

Glover, L. A., & Kelly, M. P. (1987). *Engineers in Britain - A Sociological Study of the Engineering Dimension*. Allen & Unwin publishers.

Hill, L. (1993). *Becoming a Manager: Mastery of a New Identity*. Harvard Business School Press.

Hood, D. D. (1990). Transition: Engineer to manager: Specialist to generalist. *IEEE International Engineering Management Conference*, 22-28.

Howard, C. A. (2003). *From Engineer to Engineering Manager: A qualititative study of experiences, challenges, and individual transitions for engeening managers in aerospace companies* (PhD thesis). The Pennsylvania State University.

Hsu, J. (2017). *The curious transition to becoming an engineering manager*. https://blog.gitprime.com/transition-to-engineering-manager/

Immen, W. (2010). Why most management skills can't be taught in business school. *The Globe and Mail*. https://www.theglobeandmail.com/report-on-business/careers/career-advice/why-most-management-skills-cant-be-taught-in-business-school/article1215715/

Ktoridou, D., Doukanari, E., & Epaminonda, E. (2016). Designing a product development process graduate course: An interdisciplinary example. *IEEE Xplore proceedings of IEEE EDUCON 2016 - Engineering Education towards Excellence and Innovation, IEEE Global Engineering Education Conference*, 1092-1095.

Ktoridou, D., Doukanari, E., Karayiannis, A., & Epaminonda, E. (2018). Case-based Learning: Offer-ing a Premier Targeted Learning Experience for Technology Management Students. *IEEE Xplore proceedings of IEEE EDUCON 2018 - Emerging Trends and Challenges of Engineering Education, IEEE Global Engineering Education Conference*, 1781-1786.

Ktoridou, D., & Eteokleous, N. (2012). Cultivating Entrepreneurial Skills through Case-Based Learn-ing for MIS Courses. *Proc IEEE EDUCON*, 11-15.

Lincoln, Y. S., & Guba, E. G. (1985). *Naturalistic Observation*. SAGE.

Lloyd, B.E. (1991). *Engineers Australia - A Profession in Transition*. Melbourne: Macmillan Company.

Lutchen, K. (2016). *This Is the Best Major For Every Wannabe CEO*. https://fortune.com/2016/05/22/women-stem/

Martin, R. (2010). Management Is Not a Profession — But It Can Be Taught. *Harvard Business Re-view*. https://hbr.org/2010/07/management-is-not-a-profession.htm

Martínez-León, I., Olmedo-Cifuentes, I., & Ramón-Llorens. (2018). Work, personal and cultural factors in engineers' management of their career satisfaction. *Journal of Engineering and Technology Management*, *47*, 22-36.

Mintzberg, H. (1971). Managerial Work: Analysis from Observation. *Management Science*, *18*(2), B-1–B-115. doi:10.1287/mnsc.18.2.B97

Munson, J. M., & Posner, B. Z. (n.d.). The Values of Engineers and Managing Engineers. *IEEE Transactions on Engineering Management*, *26*(4), 94–100.

National Society of Professional Engineers. (2006). *Frequently Asked Questions About Engineering*. https://web.archive.org/web/20060522214617/http://www.nspe.org/media/mr1-faqs.asp

Roberts, K., & Biddle, J. (n.d.). The Transition into Management by Scientists and Engineers - A Misallocation or Efficient Use of Human Resources? *Human Resource Management*, *33*(4).

Rottmann, C., Reeve, D., Kovalchuk, S., Klassen, M., Maljkovic, M., & Moore, E. L. (2019, June). *Counting Past Two: Engineers' Leadership Learning Trajectories*. Paper presented at 2019 ASEE Annual Conference & Exposition, Tampa, FL. https://peer.asee.org/32552

Rynes, S. L. (1987). Career Transitions from Engineering to Management: Are They Predictable among Students? *Journal of Vocational Behavior*, *30*, 138–154.

Sakakibara, K., & Westney, D. E. (1995). Comparative study of the training, careers, and organisation of engineers in the computer industry in the United States and Japan. *Hitotsubashi Journal of Commerce and Management*, *20*(I), 1–20.

Seethamraju, R., & Agrawal, R. (1999). Engineers as managers: a conceptual model of transition. In *PICMET '99: Portland International Conference on Management of Engineering and Technology. Proceedings Vol-1: Book of Summaries* (IEEE Cat. No.99CH36310). IEEE. doi: 10.1109/PICMET.1999.787818

Shepard, H. A. (1958). The dual hierarchy in research. *Research Management*, *1*, 177–187.

Souder, W. E. (1983). Planning and Career Path from Engineering to Management. *Engineering Management International*, *1*(4), 249–258.

Taylor, F. W. (2005). The Principles of Scientific Management. Harper & Row.

The Economist. (2004). *Business schools: But can you teach it?* https://www.economist.com/special-report/2004/05/20/but-can-you-teach-it

Urwick, L., & Brech, E. F. L. (1963). The makings of scientific management: vol. 1. Thirteen pioneers. Pitman.

Urwick, L. F. (1953). Management Debt to the Engineer. *Mechanical Engineering (New York, N.Y.)*, (May), 374–386.

U.S. Department of Labor. (2006). Engineers. In *Occupational Outlook Handbook, 2006–07*. Author.

Weber, M. (1978). The Technical Superiority of Bureaucratic Organisation over Administration by Notables. In G. Roth & C. Wittich (Eds.), *Economy and Society: An Outline of Interpretive Sociology* (pp. 973–975). University of California Press.

Williams, B. (1988). *Review of the Discipline of Engineering*. Australian Government Publishing Service.

Yeh. (2008). Exploring Career Stages of Midcareer and Older Engineers—When Managerial Transition Matters. *IEEE Transactions on Engineering Management, 55*(1).

Chapter 7
The Power of Leadership in Engineering Management Education

Zeyad Waleed Abualhamael
https://orcid.org/0000-0003-1915-6278
Manchester Metropolitan University, UK

EXECUTIVE SUMMARY

The aim of this chapter is to bridge the gap between literature about leadership studies and engineering education to develop future engineering leaders. The chapter focuses upon the hindrances that are faced in integrating leadership programs in the engineering circular. To this end, the chapter employed qualitative methodology under which it reviewed literature on multiple lines such as identifying significance of leadership for engineering students, identification of leadership practices and skills taught to engineering students, and the way this can be improved. Key findings of chapter emphasized on leadership as an important skill for future engineers thus demanding need for equipping engineers with skills needed to become an effective leader. It is found that having engineers to develop leadership skills will enable them to perform their role effectively, effectively allocate and utilize available resources. Thus, the chapter recommends integration of leadership knowledge and skills in engineering curricula through usage of variety of tools such as experiential learning and self-awareness.

DOI: 10.4018/978-1-7998-4063-3.ch007

INTRODUCTION

Global recognition of the "leadership" concept that has emerged over the past several years has resulted in certain challenges for various contexts in altered settings. Debates about different leadership styles and the suitability of specific approaches in various situations have been highlighted (Vroom & Jago, 2007; Bennis & Nanus, 1985). While it is evident that leadership studies have been growing significantly in many fields such as politics, organizations, health, engineering and even education (Esen, Bellibas, & Gumus, 2018), there are still a lack of consensus on the definition of leadership. While academics define leadership with adjectives and verbs (e.g. visionary, creative and empower) (Cox, Berry, & Smith, 2009), industry experts link leadership to action-oriented elements (e.g. being proactive and having good decision making skills) (Cox, Cekic, Ahn, & Zhu, 2012).

In the corporate world, engineers are frequently overlooked for leadership positions while individuals with MBAs and similar degrees are given a preference as they are better equipped with communication, leadership and management skills (Summers, Davis, & Tomovic, 2004). Several companies now demand for engineers to possess analytical and leadership skills for promotions and leadership positions (Kumar & Hsiao, 2007).

Leadership skills for engineers are important as they ensure that engineers are not only equipped with the technical knowledge but are also innovators and responsible citizens of the society, bringing about positive change. More precisely, construction companies are especially recruiting graduates with both management and leadership skills to be able to improve interpersonal skills and successfully fulfil projects (Dulaimi, 2005). According to Arciszewski (2006), lack of leadership in civil engineering was a crisis and the engineers were urged to use current challenges to evolve the profession to cater to the everchanging demands.

Understanding the wide ranging individual and organizational benefits of leadership skills, Accreditation Board for Engineering and Technology (ABET), National Academy of Engineering (NAE) and similar organizations have stressed upon a formal inclusion of leadership programs in the engineering curriculum and have associated leadership with successful recruitment and career progression (Kumar & Hsiao, 2007). Similarly, numerous undergraduate and graduate engineering programs have initiated leadership development as either part of the curricula, as a co-curricular or as an extracurricular activity.

One such example is of the Engineering Leadership Program (ELP) based on the "Engineer of 2020" vision articulated by the National Academy of Engineering in the Iowa State University. It is a 4-year program that develops leadership within the students through meaningful co-curricular activities in addition to their classroom experience (Athreya & Kalkhoff, 2010).

In the education institutions' context, the concept has begun to unleash some challenges and issues (Bryman, 2007; Scott, Coates, & Anderson, 2008), and a review of the literature reveals that there is a lack of understanding when it comes developing engineering education programs (Paul & Cowe, 2015). Furthermore, a lack of consensus on the leadership programs along with acute underrepresentation of women in the engineering profession and lower retention rates in institutions similarly undermine the effectiveness of leadership programs in terms of gender diversity.

SIGNIFICANCE OF THE CHAPTER

Globally, leadership in organizations is well known and considered a key factor that leads to either the success or failure of organizations and institutions (Trottier, Wart, & Wang, 2008; Storey, 2016). This chapter focuses on several objectives. Firstly, it aims to identify the importance of leadership studies for engineering students. Next, it focuses on providing a better and wider understanding of the adopted leadership and management studies within engineering education in different forms.

It also aims on expanding the knowledge in adopting more progressive methods in higher educational institutions to improve education standards of engineering students as well as address the leadership challenges in the engineering context (Alamri, 2011; Mawson, 2001).

Moreover, an in-depth discussion on designing and implementing effective learning experiences, along with the integration of adult learning concepts (experiential learning, self-awareness, personal reflection, and articulation of vision) can serve as a beacon and a guide for institutions when designing leadership development programs for their own universities.

This study shall also serve as a starting point for readers aiming to delve deeper into the subject. While numerous studies have been conducted on this topic, most of them only focus on a single aspect of the topic. This chapter aims at providing a holistic view to the readers while covering all aspects in considerable detail

A fundamental significance of this chapter is to bridge the gap between leadership studies and engineering education to develop motivated, inspired, prepared and positioned engineers that can direct project teams and creating a collective working environment. Eventually, the chapter will be addressing the incorporation of leadership development and engineering education programs as a respond of the global processional demands "global leaders building a better quality of life" (Mawson, 2001).

WHAT IS LEADERSHIP?

The function of teamwork in organizations have evolved into an important topic in terms of theoretical, applied and empirical research (Baker & Gerlowski, 2007). With increasing complexity in the organizational world, modern firms are more reliant than ever before on teams, which has led to a recent surge towards a more comprehensive and detailed perspective on the workings between teams and leadership (Daspit, Tillman, Boyd, & Mckee, 2013; Denis, Langley, & Sergi, 2012; Driskell, Goodwin, O'Shea, & Salas, 2006; Baker & Gerlowski, 2007).

Traditionally, a leader has been known as a singular individual with a purpose to "provide orders, delegate tasks, set expectations, provide guidance, set the rules and coordinate daily activities for the followers" (Chin, 2015). However, with the progress and global competence of organizations, there has been a shift towards diffusion of power (horizontal leadership) to leverage knowledge, information, and resources, while catering to the diversity and the multitude of personalities that comprise the teams (Gordon, 2002). This leads to the idea of distribution of power being more attractive than being concentrated to one individual (Callanan, 2004).

Scholars of diverse and various fields including social science, social psychology, industrial organization, sociology, and psychology have presented their research on leadership and its types in teams and organizations (Stewart, 2010; Sundstrom, McIntyre, Halfhill, & Richards, 2000). To ensure conciseness and to avoid running over old ground, it was deemed unnecessary to review several old leadership styles, such as the 'Trait Approach', the 'Skill Approach', 'Situational Approach', 'Contingency Approach'. This section primarily will focus upon behavioral approach to leadership and contemporary theories of leadership

Behavioral approach to leadership defines leadership with the view that leaders are responsible for shaping an environment which helps the subordinates/followers achieve their goals (Stajkovic & Luthans, 1997). This approach is important as it addresses the diversity in the organizations. Moreover, due to problems of translating theory into practice due to validity of measurement questionnaires, observational measures could help managers to attempt to lead diverse organizations (Mosley, 1998). The behavioral approach includes autocratic, democratic, Laisse-Faire and paternalistic leadership styles. To give readers a flavor of behavioral leadership style and maintain conciseness, only autocratic leadership is discussed. Details on other styles can be easily searched upon online.

Autocratic leadership, also known as authoritarian leadership, is a style dominated by control over all decisions with minimal input from others. Some primary features include little to no input from team members and subordinates, leaders make judgements based on individual intellect, and orders and methods are dictated with group members feeling untrusted with important tasks and decisions. Authoritarian style tends to create rigid and structured hierarchy structures, discourages innovation and out of the box thinking and puts forward clearly defined rules with minimal flexibility (Wang & Guan, 2018). Such a style is preferred when the leader is the most knowledgeable in the group and requires swift decisions (Schaeffer, 2002).

On the other hand, contemporary leadership theory redefined leadership as a dynamic and reciprocal process among individuals moving towards a similar goal. A contemporary leader uses personal influence to inspire people to develop and achieve organizational goals. It is unlike the traditional forms which primarily focused on the leader rather than the empowerment of the follower as well (Komives & Dugan, 2010). While contemporary approaches include transformational leadership, leader-member exchange, servant leadership, and authentic leadership, transformational leadership shall be discussed in detail as it is the leadership style that is more famous among leaders, especially women (Sellgren, Ekvall, & Tomson, 2006). For further clarity on the aspects of transformational leadership, transactional leadership is also briefly introduced for readers to compare and understand the development of this theory.

Transactional leadership is a process of "transaction" between the leader and the team members. Such leaders are committed to short team goals, follow existing procedures, ignore the need to develop ideas and innovation with focus on personal motives rather than shared interests and no explicit moral code (Ciulla, Day, & Antonakis, 2011). Transactional leadership comprises of mainly three essential components:

- Contingent reward: transaction in terms of payment and reward in exchange of good efforts and employees failing are "penalized"
- Management by exception-drive: it presumes constant guidance by the leader, pointing deviations and taking remedial action (Antonakis, 2001).
- Management by exception-passive: it presumes the leader's intervention in only extreme cases when standards are not met (Antonakis, 2001; Ciulla, Day, & Antonakis, 2011).

Bass (1990) argued that the "compromise version" of leadership (transactional) is limited to simple transactions. He thus developed a new leadership model, identifying it as transformational leadership, where a leader taps into the followers' higher needs and values, aims for the greater and long-term aim, and assumes a moral aspect to the leadership. With numerous additions and dimensions, the four primary characteristics comprising this style are:

- Charisma or idealized influence: it puts forward a sense of vision, mission, respect, and mutual trust.

- Inspirational motivation: the leader carries a vision which inspires and motivates others, influencing them to achieve goals beyond immediate ones.
- Intellectual stimulation: logic and intelligence are demanded in solving problems.
- Individualized consideration: the leader also acts as a mentor and guide, expressing personal responsibility towards the subordinates while empowering them (Antonakis, 2001; Ciulla, Day, & Antonakis, 2011).

LEADERSHIP IN ENGINEERING

According to UNESCO (2010), around 2.5 million new engineers and technicians would have been needed in the Sub-Saharan desert only if the UN MDG of improved access to clean water and sanitation had to be achieved. The same report also predicted a severe shortage of engineers in Germany and the UK with additional 2,217,500 employees being needed in principal engineering-related skill areas over the next ten years. Korn Ferry (2016) also notes a gradual increase in the engineering industry with 15% to 20% growth in management functions, professional engineers, and qualified technicians. Filling such positions demands possession of management and leadership skills, beyond basic technical ones.

In addition to core technical knowledge, professional skills especially leadership and team effort are being considered integral to enable individuals to be well-rounded professionals who are not only professions but drivers to innovation in the everchanging society (Cox, Cekic, & Adams, 2010). Recently, leadership is specifically promoted as minors, formal graduate and undergraduate programs, and graduate courses in engineering curriculum as a skill that shall help future engineers (Cox, Berry, & Smith, 2009).

Not only the personal self, influential leadership impacts other individuals, colleagues, and the organization on the whole, as a research understanding the employees' performance in Telecom Engineering companies by Sougui and other fellow researchers (2015) suggested. Another independent study pointed out that modern day engineering project managers and team members face ever-increasing demands. This is not only due to the intense workload of the projects, but also due to the rapidly changing business dynamics. The effect of such pressure materializes in the form of stress (Haynes & Love, 2004). This could lead to high turnover, increased absences due to illnesses, decreased job satisfaction, and monetary losses. Another recent study that looked upon the impact of stress on construction project managers (Leung, Chan, & Olomolaiye, 2008) reached similar conclusions.

Effective leadership, in terms of coaching and optimum leadership style, could be used to reduce stress levels within the engineering organization and ensure employee satisfaction (Kombarakaran, Yang, Fernandes, & Baker, 2008; Hackman & Wageman, 2005). Leadership style, as mentioned above includes emotional intelligence, self-management and positive talk, feedback, clear goals and visualization among others (Berg & Karlsen, 2013).

Development of leadership skill, studies of famous leaders and their abilities to lead go back to several centuries. However, this had been primarily focused upon by business students and social sciences students (Bolman & Deal, 2017; Morrison, 2007). However, in the recent past, many organizations including National Academy of Engineering (2004), National Research Council and Accreditation Board for Engineering and Technology (2018) have played an active part in redesigning the engineering curricula and presenting possible suggestions to improve the general standard of engineers; an important trait in the career progression and responsible citizenship of the individuals. The details are presented later in the chapter. All three organizations have criticized higher education institutions for their lack of success to turn recent engineering graduates into exemplary leaders.

Currently, there is a stark disparity between the skills of current engineers graduating from institutions and the industry's need of individuals who can effectively lead teams and execute engineering projects in a cross functional environment by achieving performance targets on time using the required budget to meet stakeholders' needs successfully. This gap can only be bridged using innovative integrative approach that not only provides technical depth but also focuses on interpersonal skills that are essential to lead engineering teams successfully (Pitts, Klosterman, & McGonagle, 2013).

With the importance that leadership skills have for engineers on both individual and organizational levels, it is only fitting that harnessing and developing such skills should be made a priority in educational institutes and courses be included in the curricula to develop more well-rounded engineers.

PLANNING LEADERSHIP SKILLS IN ENGINEERING EDUCATION

Numerous leadership development programs have proven to show improvement in, and positively impact both individual and organizational performance (Virakul & McLean, 2012; Riggio, 2008). One way to incorporate leadership skills in engineering education is to formally harness and develop such skills in individuals. Numerous organizations have been working in this regard. Accreditation Board for Engineering and Technology (ABET) (2018) in its '2019-2020 Criteria for Accrediting Engineering Programs' requires the curriculum for Engineering Management and similar engineering programs to *"prepare graduates to understand the engineering relationships between the management tasks of planning, organization, leadership, control, and the human element in production, research, and service organizations"*

Likewise, other programs including Construction and Civil Engineering Programs require inclusion of communication, leadership, and decision and optimization methods by ABET as well.

Another such organization, National Academy of Engineering (NAE) (2004), in its report 'The Engineer of 2020: Visions of Engineering in the New Century' tried to prepare the future graduates of engineering universities with a simple question in mind "What will or should engineering be like in 2020?" considering the prior and evolutionary role of engineers. Chapter 4 of the said report pointed out at skills and attributes that a graduate engineer of the batch 2020 should possess. Attributes such as "strong analytical skills, creativity, ingenuity, professionalism, and leadership" were emphasized upon showing how organizations have been working tirelessly to seamlessly integrate leadership in the formal engineering curriculum.

It is also imperative that an environment that facilitates and nurtures learning be created by the facilitator (Hoover, Giambatista, Sorenson, & Bommer, 2010). It is also important for the learner to be willing as the learning process depends upon the intention of the learner to learn (Vohra, Rathi, & Bhatnagar, 2015).

There are several models that describe and provide an insight to designing and implementing effective learning experiences. Behaviorism focuses on a changing stimulus to attain a variety of responses and thus train and nurture the desired response. Cognitivism stresses upon reinforcing positive behaviors and actions and discarding the negative ones. Repetition of actions occurs through gaining new knowledge and applying it in real life. Social theorists focus upon the learner interacting with the environment and learning by forming a relationship with it. According to transformative learning, learners learn through introspection and critical reflection of the environment and its proponents around the individual (Joplin, 1981; Luckmann, 1996; Hornyak & Page, 2004; Vohra, Rathi, & Bhatnagar, 2015). According to a study conducted by Allen in 2007 (p. 36), "leadership development program that incorporates the thinking of behaviorists, cognitivists, social learning theorists, and developmentalists will not only involve learners at a higher level, it will help architects of leadership development programming design interventions and environments more conducive to learning". Therefore, inculcating leadership programs in engineering education requires facilitators to develop theories that integrate the adult learning concepts.

Numerous researchers have concluded that experiential training is very effective in developing leadership and team building skills (Hoover, Giambatista, Sorenson, & Bommer, 2010; Judge, 2005; Shivers-Blackwell, 2004; Sibthorp, Paisley, &

Gookin, 2007). In such a training, the facilitator takes the learner through a series of physical experiences (activities) with reflection upon the activities being highly encouraged. This helps the participants form valuable insights and transfer them to a work context (Vohra, Rathi, & Bhatnagar, 2015). Experiential learning offers several significant advantages over classroom teaching. For instance, physical involvement encourages active learning. Similarly, experiential learning helps in the development of self-concept, improves communication and problem solving, and encourages individual risk taking (Hornyak & Page, 2004; Shivers-Blackwell, 2004; Sibthorp, Paisley, & Gookin, 2007; Watson & Vasilieva, 2007; Williams, Graham, & Baker, 2003). Participation promotes problem solving, trust, and experiencing real life emotions which form real life judgement among participants. This is something that a traditional classroom, with individual feedbacks, isolated learning and classroom instruction, may be unable to provide (Judge, 2005).

Another observation explains that successful leaders portray high levels of self-awareness (Vohra, Rathi, & Bhatnagar, 2015). Developing self-awareness for an individual entail developing a keen eyesight on the motives, attitude, behavior, feedback, willingness to search for developmental opportunities and feedback (Riggio, 2008; Avolio & Hannah, 2008; Thompson, Grahek, Phillips, & Fay, 2008). Such individuals transform into leaders that realize their strengths and weaknesses, find it comfortable discussing them and are ready to listen to constructive criticism (Vohra, Rathi, & Bhatnagar, 2015). For further character development, leaders need to be open to feedback, listen to advices and self-critique their leadership based on reviews from superiors, peers and subordinates (Riggio, 2008).

Apart from experiential learning and self-awareness, leaders also need to learn how to reflect on their experiences both in personal and professional life. Avolio & Hannah (2008) emphasize that a leader's life is made of experiences which teach him important lessons and play an integral role in leadership development. "By constructing one's narrative or life story, people proactively make meaning of their experiences, thereby forming their identity or self-construct which then becomes the lens through which one views future experiences" (Vohra, Rathi, & Bhatnagar, 2015). In addition to learning from life experiences, it is also important for a leader to learn through simulated learning which helps one streamline his thought process, transfer knowledge from simulations to real life and enhance learning (Amagoh, 2009).

Boyatzis (2009) stresses that writing a personal vision for the future and sharing it with trusted people nurtures leadership. Not only does this reiterate and clarify the goals in life, it also helps the individual articulate his ideal self.

While numerous workshops have been planned on such principles for varying disciples across the globe, an interesting workshop conducted by Vohra and colleagues (2015) for Executive MBA (EMBA) students encompasses all the leadership theories explained above and is an ideal workshop for implementation. While used with

Business Management students, the workshop has the ability to be replicated for different disciplines and hopes to initiate a dialogue among faculty members and instructors.

An outdoor two-and-a-half-day workshop with the ability to cater large groups and integrate mixed ideologies, the workshop, and the reflections immediately after the tasks and a month later, strengthened their learning from the workshop. The table below shows the design of the workshop, description of activities, learning methodologies and objectives:

Table 1. Design of the Workshop and Description of the Activities, Learning Methodology, and Learning Objectives

	Activities and Their Brief Description	Learning Methodology	Learning Objectives	Time Allotted
1	**Marble Transport:** Transport marbles into a bowl, using pipes, following certain rules within specified time.	Outdoor experiential activity.	• Align interests. • Observe patterns.	1.5 Hour
2	**Making the Square:** make a perfect square, using the rope provided, following certain rules within specified time.	Outdoor experiential activity.	• Foster trust. • Align interest. • Openness.	1.5 Hour
3	**Adventure Activities:** participate in two of the three activities, namely rock climbing, rappelling and river crossing.	Outdoor experiential activity.	• Motivate self and others. • Take risks. • Mentor and help in development of others.	3 Hour
4	**Drawing Fractals:** Drawing significant memories since childhood. Sharing life story with another person and guided analysis of life story.	Reflecting on life experiences and self-awareness.	• Develop self-awareness and self-exploration. • Foster trust. • Encourage openness. • Observe patterns.	8 Hour
5	**Personal Vision:** Write a vision statement for oneself.	Envisioning and reflection.	• Develop a personal vision.	2 Hour
6	**Nature Walk:** Take a walk in the open area nearby, observe the surroundings, and draw lessons for leadership from this observation.	Reflection.	• Be open to new ideas. • Learn continuously.	1 Hour
7	**Feedback Giving and Receiving:** Seek and provide genuine and constructive feedback with the intention to help and grow.	Seeking feedback.	• Provide and receive feedback. • Enhance self-awareness. • Provide opportunity to look for areas of improvement. • Impart insights on how to give feedback to others.	2 Hour

Source: (Vohra, Rathi, & Bhatnagar, 2015)

Moving on from extra-curricular activities to co-curricular activities, it is proven that holding leadership positions in engineering clubs, student chapters of a professional society, engineering clubs especially focused on minority and female students, and non-engineering clubs have consistently reported to improve leadership, communication and teamwork skills (Carter Et Al., 2016). This is in line with the methodologies mentioned above of experiential learning, self-awareness and etcetera.

This serves a beacon for institutions to promote societies and extra-curricular activities in universities. Greater budget allocation, encouragement for students, monetary and non-monetary prizes in competitions shall further encourage participation. Also, after the activities, the students could be asked to present a reflection paper to better remember the insights gained from those activities.

While experiential learning is preferred over classroom learning as mentioned above, class alignment with physical activities and the industry requirement could further aide learning. Incorporating numerous leadership theories mentioned previously in this chapter, Professor Pitts and his colleagues (2013) discuss the Gordon Engineering Leadership Program (GEL), a transformational graduate program in the Northeastern University's College of Engineering. The program first teaches relevant leadership theories which are followed by practice in leadership laboratories. Technical product development and scientific foundation then lead to a market-worthy challenge project. A three-way mentoring program including mentors from the industry, faculty and program mentors supplement the program.

The foundation of the course encompasses of four elements which are integrated into every aspect of the program:

- Experiential learning to enhance knowledge, behavior, and skills.
- Guest speaker sessions with distinguished individuals to discuss and model leadership.
- The three-way mentoring programs.
- Cross-cohort sharing where candidates broaden their perspectives through each other.

Five pillars represent the core of the program:

- Leadership capabilities
- Leadership laboratories
- Product development
- Scientific foundations
- Challenge project

The foundational and the core elements work together to develop both technical and human capabilities of engineering leadership in the individuals. The primary aim of this course is to satisfy the market, the customer, and the stakeholders.

The program has been successful with industry personnel, and candidates endorsing the methods and the design of the program. According to one individual:

"The leadership labs and activities set a strong tone for how I should act and behave as a leader. At the end of this program, not only did I learn the class material, but I have grown. I have become more confident and proactive"

CHALLENGES OF INTEGRATING THE LEADERSHIP SKILLS IN THE ENGINEERING CONTEXT

While leadership development offers a variety of advantages in terms of both individual and organizational levels, there are several challenges for successfully implementing and integrating leadership skills in the engineering context. Such challenges are of a contextual nature and require unique solutions. Others are general problems that tend to exist across disciplines.

Smithers & Walker (2000) point out that women are highly under-represented in engineering and women face barriers to attain positional leadership despite numerous political and educational initiatives (Hoyt & Blascovich, 2007; Kark & Eagly, 2010). However, this unjust behavior has its roots firmly planted in educational institutes that are supposed to nurture individuals irrespective of gender (Dugan, Fath, Howes, Lavelle, & Polanin, 2013). Such an inequality plays a detrimental role in restricting women's role, career and persistence in the STEM fields (Rosser, 2004). This underrepresentation manifests in terms of lower education retention rates in STEM in comparison to all other disciples (Hill, Corbett, & Rose, 2010). Another manifestation of this inequality are reduced levels of leader efficacy (one's internal belief in the ability to perform leadership related tasks (Hannah, Avolio, Luthans, & Harms, 2008)) in female STEM majors in comparison to their non-STEM counterparts.

Next, Undergraduate Research (UR), another important aspect for the engineers during their bachelors, has had minimal research performed upon. Most researches that have attempted to link leadership to undergraduate research have had limited scope and causal conclusions due to methodological limitations and smaller sample sizes (Seymour, Hunter, Laursen, & Deantoni, 2004). Moreover, researchers in engineering education have focused upon leadership as a composite outcome along with teamwork and communication rather than studying leadership as a discreet concept (Sankar & Raju, 2011; Seat, Parsons, & Poppen, 2001; Tonso, 2006).

However, it is argued by Carter and other fellow researchers (2016) that additional skills, such as planning and motivating others, are required to learn and develop leadership which shows that researchers need better analysis to explain leadership.

Moreover, the basic nature of the leadership program is still hazy with many researchers not reaching a common ground. A great variation in leadership programs (Brungardt, Greenleaf, Brungardt, & Arensdorf, 2006; Jenkins & Owen, 2016; Seemiller & Murray, 2013), despite having a similar name, lead towards completely different outcomes which question the validity and credibility of the field itself (Whitaker & Arensdorf, 2018). While discussing the diversity of perspectives in the framework of the basic elements of the field, Harvey & Riggio (2011) stated "… not only do we have different disciplinary points of view, but in describing leadership…, we are speaking different languages" (Harvey, 2011, p. 4).

CONCLUSION

This chapter has emphasized on leadership as an important skill for the engineers to possess. The evolving marketplace demands engineers to not only possess technical knowledge, but also develop other highly important skills of teamwork and leadership. Effective leadership not only enables engineers to become well-rounded individuals, it also positively impacts colleagues and the organization. Effective leadership also serves as a stress reliever for project managers who have demands to meet and clients to satisfy.

Therefore, leadership needs to be integrated into the engineering curricula. The integration could be through experiential learning, self-awareness, personal reflection, and articulation of vision. Activities incorporating these theories shall be optimum for learners to harness leadership.

However, the traditionalism in the engineering field and curricula hampering growth of women in the field, lack of clarity of the fundamentals of leadership and the composite nature of leadership in various studies have led to challenges in integration of leadership in engineering.

Conclusively, the irrefutable importance of leadership in engineering and its numerous advantages for engineers stress that it be integrated into engineering on an urgent basis and the process of change needs to be inclusive to reap maximum rewards. Mentorship and personal coaching could be helpful in developing leaders as responsible citizens and success individuals.

A limitation of this chapter is the relative lack of depth for the different aspects covered. While trying to provide a holistic view to the readers with a reasonable length, additional details had to be cut out to ensure that the interest of the reader does not wane. Future researchers could use this chapter as a starting point to build their base, gain clarity, understand the basic concepts and then dig deeper into the plethora of literature available on this topic.

REFERENCES

Accreditation Board for Engineering and Technology (ABET). (2018). *Criteria for accrediting engineering programs*. ABET.

Alamri, M. (2011). Higher Education in Saudi Arabia. *Journal of Higher Education Theory and Practice, 11*(4), 88–91.

Allen, S. J. (2007). Adult learning theory and leadership development. *Leadership Review, 7*, 26–37.

Amagoh, F. (2009). Leadership development and leadership. *Management Decision, 47*(6), 989–999. doi:10.1108/00251740910966695

Antonakis, J. (2001). *The validity of the transformational, transactional, and laissez-faire leadership model as measured by the Multifactor Leadership Questionnaire (MLQ 5X)* (Unpublished PhD thesis). Minneapolis, MN: Walden University ScholarWorks.

Arciszewski, T. (2006). Civil engineering crisis. *Leadership and Management in Engineering, 6*(1), 26–30. doi:10.1061/(ASCE)1532-6748(2006)6:1(26)

Athreya, K. S., & Kalkhoff, M. T. (2010). The Engineering Leadership Program: A co-curricular learning environment by and for students. *Journal of STEM Education, 11*(3-4), 70-74.

Avolio, B. J., & Hannah, S. T. (2008). Developmental readiness: Accelerating leader development. *Consulting Psychology Journal: Practice and Research, 60*(4), 331–347. doi:10.1037/1065-9293.60.4.331

Baker, S., & Gerlowski, D. (2007). Team effectiveness and leader-follower agreement: An empirical study. *The Journal of American Academy of Business, Cambridge, 12*(1), 15–23.

Bass, B. M. (1990). From Transactional to Transformational Leadership: Learning to Share the Vision. *Organizational Dynamics, 18*(3), 19–32. doi:10.1016/0090-2616(90)90061-S

Bennis, W., & Nanus, B. (1985). Strategies for Taking Charge. In *Leaders*. Harper.

Berg, M. E., & Karlsen, J. T. (2013). Managing Stress in Projects Using Coaching Leadership Tools. *Engineering Management Journal*, *25*(4), 52–61. doi:10.1080/10429247.2013.11431995

Bolman, L. G., & Deal, T. E. (2017). *Reframing Organizations: Artistry, choice, and leadership* (6th ed.). John Wiley and Sons, Inc. doi:10.1002/9781119281856

Boyatzis, R. E. (2009). Leadership development from a complexity perspective. *Consulting Psychology Journal: Practice and Research*, *60*(4), 298–313. doi:10.1037/1065-9293.60.4.298

Brungardt, C., Greenleaf, J., Brungardt, C., & Arensdorf, J. (2006). Majoring in leadership: A review of undergraduate leadership degree programs. *Journal of Leadership Education*, *5*(1), 4–25. doi:10.12806/V5/I1/RF1

Bryman, A. (2007). Effective leadership in higher education: A literature review. *Studies in Higher Education*, *32*(6), 693–710. doi:10.1080/03075070701685114

Callanan, G. A. (2004). What would Machiavelli think? An overview of the leadership challenges in team-based structures. *Team Performance Management*, *10*(3/4), 77–83. doi:10.1108/13527590410545072

Carter, D. F., Ro, H. K., Alcott, B., & Lattuca, L. R. (2016). Co-Curricular Connections: The Role of Undergraduate Research Experiencess in Promoting Engineering Students' Communication, Teamwork, and Leadership Skills. *Research in Higher Education*, *57*(3), 363–393. doi:10.100711162-015-9386-7

Chin, R. J. (2015). Examining teamwork and leadership in the fields of public administration, leadership, and management. *Team Performance Management*, *21*(3/4), 199–216. doi:10.1108/TPM-07-2014-0037

Ciulla, J. B., Day, D. D., & Antonakis, J. (2011). *Ethics Effectiveness: The Nature of Good Leadership*. Sage Publications, Inc.

Cox, M. F., Berry, C. A., & Smith, K. A. (2009). Development of a Leadership, Policy, and Change Course for Science, Technology, Engineering, and Mathematics Graduate Students. *Journal of STEM Education, 10*(3-4), 9-16.

Cox, M. F., Cekic, O., & Adams, S. G. (2010). Developing Leadership Skills of Undergraduate Engineering Students: Perspectives from engineering faculty. *Journal of STEM Education, 11*(3-4), 22-33.

Cox, M. F., Cekic, O., Ahn, B., & Zhu, J. (2012). Engineering professionals' expectations of undergraduate engineering students. *Leadership and Management in Engineering*, *12*(2), 60–70. doi:10.1061/(ASCE)LM.1943-5630.0000173

Daspit, J., Tillman, C. J., Boyd, N. G., & Mckee, V. (2013). Cross-functional team effectiveness: An examination of internal team environment, shared leadership and cohesion influences. *Team Performance Management*, *19*(1/2), 32–56. doi:10.1108/13527591311312088

Denis, J.-L., Langley, A., & Sergi, V. (2012). Leadership in the Pluran. *The Academy of Management Annals*, *6*(1), 211–283. doi:10.5465/19416520.2012.667612

Driskell, J. E., Goodwin, G. F., O'Shea, P. G., & Salas, E. (2006). What Makes a Good Team Player? Personality and Team Effectiveness. *Group Dynamics*, *10*(4), 249–271. doi:10.1037/1089-2699.10.4.249

Dugan, J. P., Fath, K. Q., Howes, S. D., Lavelle, K. R., & Polanin, J. R. (2013). Developing the Leadership Capacity and Leader Efficacy of College Women in Science, Technology and Math fields. *The Journal of Leadership Studies*, *7*(3), 6–23. doi:10.1002/jls.21292

Dulaimi, M. F. (2005). The influence of academic education and formal training on the project manager's behavior. *Journal of Construction Research*, *6*(1), 179–193. doi:10.1142/S1609945105000328

Esen, M., Bellibas, M. S., & Gumus, S. (2018). The Evolution of Leadership Research in Higher Education for Two Decades (1995-2014): A Bibliometric and Content Analysis. *International Journal of Leadership in Education*, *23*(3), 259–273. doi: 10.1080/13603124.2018.1508753

Gordon, J. (2002). A perspective on team building. *The Journal of American Academy of Business, Cambridge*, *2*(1), 185–188.

Hackman, J. R., & Wageman, R. (2005). A Theory of Team Coaching. *Academy of Management Review*, *30*(2), 269–287. doi:10.5465/amr.2005.16387885

Hannah, S. T., Avolio, B., Luthans, F., & Harms, P. D. (2008). Leadership efficacy: Review and future directions. *The Leadership Quarterly*, *19*(6), 669–692. doi:10.1016/j.leaqua.2008.09.007

Harvey, M., & Riggio, R. E. (2011). *Leadership studies: The dialogue of disciplines*. Edward Elgar Publishing. doi:10.4337/9780857936486

Haynes, N. S., & Love, P. E. (2004). Psychological adjustment and coping among construction project managers. *Construction Management and Economics*, *22*(2), 129–140. doi:10.1080/0144619042000201330

Hill, C., Corbett, C., & Rose, A. S. (2010). *Why so few? Women in science, technology, engineering, and mathematics*. AAUW.

Hoover, J. D., Giambatista, R. C., Sorenson, R., & Bommer, W. H. (2010). Assessing the Effectiveness of Whole Person Learning Pedagogy in Skill Acquisition. *Academy of Management Learning & Education*, *9*(2), 192–203.

Hornyak, M. J., & Page, D. (2004). Experiential learning: Introducing faculty and staff to a university leadership development program. *Simulation & Gaming*, *35*(4), 461–475. doi:10.1177/1046878104263974

Hoyt, C. L., & Blascovich, J. (2007). Leadership efficacy and Women Leaders' Responses to Stereotype Activation. *Group Processes & Intergroup Relations*, *10*(4), 595–616. doi:10.1177/1368430207084718

Jenkins, D. M., & Owen, J. E. (2016). Who teaches leadership? A comparative analysis of faculty and student affairs leadership educators and implications for leadership learning. *Journal of Leadership Education*, *15*(2), 98–113. doi:10.12806/V15/I2/R1

Joplin, L. (1981). On defining experiential education. *Journal of Experiential Education*, *4*(1), 16–20. doi:10.1177/105382598100400104

Judge, W. (2005). Adventures in creating an outdoor leadership challenge course for an EMBA program. *Journal of Management Education*, *29*(2), 284–300. doi:10.1177/1052562904265251

Kark, R., & Eagly, A. H. (2010). Gender and leadership: Negotiating the labyrinth. In J. C. Chrisler & D. R. McCreary (Eds.), *In Handbook of gender research in psychology* (pp. 443–468). Springer. doi:10.1007/978-1-4419-1467-5_19

Kombarakaran, F. A., Yang, J. A., Fernandes, P. B., & Baker, M. N. (2008). EXECUTIVE COACHING: IT WORKS! *Consulting Psychology Journal: Practice and Research*, *60*(1), 78–90. doi:10.1037/1065-9293.60.1.78

Komives, S., & Dugan, J. (2010). Contemporary leadership theories. In R. Cuoto (Ed.), *Political and civic leadership: A reference handbook* (pp. 111–120). Sage. doi:10.4135/9781412979337.n14

Korn Ferry. (2016). *Tomorrow's engineering workforce*. Academic Press.

Kumar, S., & Hsiao, J. K. (2007). Engineers Learn "Soft Skills the Hard Way": Planting a Seed of Leadership in Engineering Classes. *Leadership and Management in Engineering*, *7*(1), 18–23. doi:10.1061/(ASCE)1532-6748(2007)7:1(18)

Leung, M.-y., Chan, Y.-S., & Olomolaiye, P. (2008). Impact of Stress on the Performance of Construction Project Managers. *Journal of Construction Engineering and Management*, *13*(8), 644–652. doi:10.1061/(ASCE)0733-9364(2008)134:8(644)

Luckmann, C. (1996). Defining Experiential Education. *Journal of Experiential Education, 19*(1), 6–7. doi:10.1177/105382599601900101

Mawson, T. (2001). Ready! aim! inspire! Leadership in engineering. *Leadership and Management in Engineering, 1*(2), 50–51. doi:10.1061/(ASCE)1532-6748(2001)1:2(50)

Morrison, J. L. (2007). Book Review: Northouse, Peter G. Leadership: Theory and Practice. *Journal of Education for Business, 83*(2), 116–117. doi:10.3200/JOEB.83.2.116-120

Mosley, A. L. (1998). A Behavioral Approach to Leadership: Implications for Diversity in Today's Organizations. *The Journal of Leadership Studies, 5*(1), 39–50. doi:10.1177/107179199800500105

National Academy of Engineering (NAE). (2004). *The Engineer of 2020: Visions of Engineering in the New Century*. The National Academies Press.

Paul, R., & Cowe, L. F. (2015). Engineering leadership education: A review of best practices. *Conference on Postsecondary Learning and Teaching*. University of Calgary. 10.18260/p.23972

Pitts, S., Klosterman, S., & McGonagle, S. (2013). A successful approach to educating engineering leaders at the graduate level. *Proceedings of the Canadian Engineering Education Association (CEEA)*. 10.24908/pceea.v0i0.4819

Riggio, R. E. (2008). Leadership development: The current state and future expectations. *Consulting Psychology Journal: Practice and Research, 60*(4), 383–392. doi:10.1037/1065-9293.60.4.383

Rosser, S. V. (2004). *The science glass ceiling: Academic women scientists and the struggle to succeed*. Routledge.

Sankar, C. S., & Raju, P. K. (2011). Use of presage-pedagogy-process-product model to assess the effectiveness of case study methodology in achieving learning outcomes. *Journal of STEM Education: Innovations & Research, 12*(7), 45–56.

Schaeffer, L. D. (2002). The Leadership Journey. *Harvard Business Review*, 42–47. PMID:12389460

Scott, G., Coates, H., & Anderson, M. (2008). *Learning leaders in times of change: Academic leadership capabilities for Australian higher education*. University of Western Sydney and Australian Council for Educational Research.

Seat, E., Parsons, J. R., & Poppen, W. A. (2001). Enabling engineering performance skills: A program to teach communication, leadership, and teamwork. *Journal of Engineering Education*, *90*(1), 7–12. doi:10.1002/j.2168-9830.2001.tb00561.x

Seemiller, C., & Murray, T. (2013). The common language of leadership. *The Journal of Leadership Studies*, *7*(1), 33–45. doi:10.1002/jls.21277

Sellgren, S., Ekvall, G., & Tomson, G. (2006). Leadership styles in nursing management: Preferred and perceived. *Journal of Nursing Management*, *14*(5), 348–355. doi:10.1111/j.1365-2934.2006.00624.x PMID:16787469

Seymour, E., Hunter, A.-B., Laursen, S. L., & Deantoni, T. (2004). Establishing the Benefitsof Research Experiencesfor Undergraduates in the Sciences: First Findings from a Three-Year Study. *Science Education*, *88*(4), 493–534. doi:10.1002ce.10131

Shivers-Blackwell, S. L. (2004). Reactions to outdoor teambuilding initiatives in MBA education. *Journal of Management Development*, *23*(7), 614–630. doi:10.1108/02621710410546632

Sibthorp, J., Paisley, K., & Gookin, J. (2007). Exploring Participant Development Through Adventure-Based Programming: A Model from the National Outdoor Leadership School. *Leisure Sciences*, *29*(1), 1–8. doi:10.1080/01490400600851346

Smithers, G. L., & Walker, D. H. (2000). The effect of the workplace on motivation and demotivation of construction professionals. *Construction Management and Economics*, *18*(7), 833–841. doi:10.1080/014461900433113

Sougui, A. O., Bon, A. T., & Hassan, H. M. (2015). The Impact of Leadership Styles on Employees' performance in Telecom Engineering companies. *Australian Journal of Basic and Applied Sciences*, *8*(4).

Stajkovic, A. D., & Luthans, F. (1997). A meta-analysis of the effects of organizational behavior modification on task performance, 1975-95. *Academy of Management Journal*, *40*(5), 1122–1149.

Stewart, G. L. (2010). The past twenty years: Teams research is alive and well at the journal of management. *Journal of Management*, *36*(4), 801–805. doi:10.1177/0149206310371512

Storey, J. e. (2016). *Leadership in Organizations: Current issues and key trends* (3rd ed.). Routledge. doi:10.4324/9781315695792

Summers, M., Davis, B., & Tomovic, C. (2004). When engineering and technology skills are not enough: Engineering leaders out of their element. *Proceedings of the CIEC Conference*.

Sundstrom, E., McIntyre, M., Halfhill, T., & Richards, H. (2000). Work groups: From the hawthorne studies to work teams of the 1990s and beyond. *Group Dynamics*, *4*(1), 44–67. doi:10.1037/1089-2699.4.1.44

Thompson, A. D., Grahek, M., Phillips, R. E., & Fay, C. L. (2008). The Search for Worthy Leadership. *Consulting Psychology Journal: Practice and Research*, *60*(4), 366–382. doi:10.1037/1065-9293.60.4.366

Tonso, K. L. (2006). Teams that work: Campus culture, engineer identity, and social interactions. *Journal of Engineering Education*, *95*(1), 25–37. doi:10.1002/j.2168-9830.2006.tb00875.x

Trottier, T., Wart, M. V., & Wang, X. (2008). Examining the Nature and Significance of Leadership in Government Organizations. *Public Administration Review*, *68*(2), 319–333. doi:10.1111/j.1540-6210.2007.00865.x

UNESCO. (2010). *Engineering: Issues Challenges and Opportunities for Development*. UNESCO Publishing.

Virakul, B., & McLean, G. N. (2012). Leadership development in selected leading Thai companies. *The Journal of Leadership Studies*, *6*(1), 6–22. doi:10.1002/jls.21224

Vohra, N., Rathi, N., & Bhatnagar, D. (2015). Developing Leadership Skills among EMBA Students: Innovations in Design. *The Journal for Decision Makers*, *40*(1), 15–27. doi:10.1177/0256090915573612

Vroom, V. H., & Jago, A. G. (2007). The Role of the Situation in Leadership. *The American Psychologist*, *62*(1), 17–24. doi:10.1037/0003-066X.62.1.17 PMID:17209676

Wang, H., & Guan, B. (2018). The Positive Effect of Authoritarian Leadership on Employee Performance: The Moderating Role of Power Distance. *Frontiers in Psychology*, *9*, 357. doi:10.3389/fpsyg.2018.00357 PMID:29628902

Watson, S., & Vasilieva, E. (2007). Wilderness thinking: A novel approach to leadership development. *Industrial and Commercial Training*, *39*(5), 242–245. doi:10.1108/00197850710761927

Whitaker, B., & Arensdorf, J. (2018). Big Question for Leadership Education. *The Journal of Leadership Studies*, *12*(2), 46–48. doi:10.1002/jls.21575

Williams, S. D., Graham, T. S., & Baker, B. (2003). Evaluating outdoor experiential training for leadership and team building. *Journal of Management Development*, *22*(1), 45–59. doi:10.1108/02621710310454851

Chapter 8
Management Education, Satisfaction, and Career Growth:
A Perspective of Techno-Managers

Venugopal Merugu
National Institute of Technology, Warangal, India

Rahul Thangeda
National Institute of Technology, Warangal, India

EXECUTIVE SUMMARY

The chapter aims at addressing some of the critical questions that will be of immense help to engineering graduates who are likely to consider pursuing management degrees. Data were obtained from a sample of 96 techno managers to explore their perception of career growth and opportunities after pursuing their management education. In particular, the chapter examines in detail the relationship between management education and career satisfaction in techno managers. Also, factors that influence in choosing a management degree were studied. A few of the critical findings from the study were that gender has not shown any significant difference in satisfaction. At the same time, experience has shown a significant difference in satisfaction. From the results obtained, it could be concluded that the techno managers were found to be satisfied with the career choice as an MBA after engineering.

DOI: 10.4018/978-1-7998-4063-3.ch008

INTRODUCTION

The world is moving at a faster pace. In this Volatility, Uncertainty, Complexity, and Ambiguity (VUCA) world, there is a considerable need for leaders and efficient individuals who can face challenges and turbulence in their roles at their organizations. In most cases, an engineer in today's world is technically strong in his specialization and is transforming himself into a manager. While this transformation happens, it would be of immense use to have pursued a management degree. A management degree will help engineers to make their final transformation into an all-rounder, charged with looking after a variety of departments and functions, not just their technical teams. It paves the way for accelerated promotion and enables them to realize their potential at an earlier age of their career.

Engineers are accustomed to solving technical problems, but managerial situations often require the use of much broader thinking, involving, and having concepts like financial or marketing or human resources. As they advance to the top-level positions in their career, they will need to think beyond their engineering domain, i.e., overall business like working with teams, projects, budgets, new product development, understanding customer needs, and analyzing market dynamics. These skills are essential as skills of engineer viz. elements of design, development, and engineering production (Caldecote, 1986). One who understands every aspect of business and builds a robust set of business skills can perform their duties efficiently and effectively. Management skills can take an engineer's career to the next level. To stay competitive in the job market, aspiring managers and engineers who want to advance in their careers need a diverse and robust set of skills. Baruch and Peiperl (2000) found MBA graduates to report higher workplace skills/competencies after their MBA education.

In India, the engineering curriculum represents more about their specialization in their engineering stream. After understanding the need for the management essentials in engineers, some of the technical institutions have started introducing one or two courses like engineering economics, accountancy, project management, human resource management, marketing management, and entrepreneurship in their curriculum. It is only when their roles begin to change, and while they grow up their career and when they start to take on broader responsibilities, they will realize the importance of these skills. Studying one or two courses in management essentials may not help an engineer to work efficiently or to face the problems that arise in the real world. To overcome them, engineers would need to have the management blend. A successful career after MBA entails both securing the entry-level position and doing well enough in both that position and subsequent positions to have a career growth.

Many engineering students would want to turn themselves into an entrepreneur by seeking entry to the world with their innovative ideas. They might have a very creative idea, but it should also be supported by awareness in business decision-making to have a strategic advantage in the current competitive world. The one who understands both technology and its social potentials has a competitive edge on the problems she faces (Ben-Haim, 2000).

Progressing to a higher level in the organization structure is the motive behind choosing the course. Thus, it is essential to know how vital an MBA is after engineering. Also, it will be of use to study about pursuing an MBA program after engineering makes a positive contribution to career success and satisfaction. Management education is vital for engineers in a competitive world. If an MBA does not contribute to career success and satisfaction, perhaps management institutes should reconsider their curriculum and teaching methods. It is evident and is need of the hour to study the factors contributing to the career satisfaction of the MBA after engineering. Thus it will help lots of prospective learners decide on choosing an MBA program. Given this background, understanding possible influencing factors on the choice of student's career is an essential issue for institutions. More importantly, how satisfied with the profession as well as job need to be addressed so that upcoming techno managers will decide to have a career in MBA. This chapter aims at throwing some light on the relationship between MBA education and the subsequent upward mobility in management that is career satisfaction. In other words, does being a techno-manager enhance long-run success is the critical question that this chapter aims at addressing.

The chapter aims at the understanding perception of the respondents towards the MBA programme. Efforts are made to highlight some of the critical aspects like what factors motivated to pursue MBA and at what state did they decide to pursue. The focal point of the chapter is to investigate the techno manager's career satisfaction and job satisfaction. Also, respondents recommending prospects to follow MBA are studied to give the prospects a fair amount of insights to enable them to make decisions.

The study addresses the following research questions:

RQ1: At what stage did the respondents decide to do an MBA program?
RQ2: What factors motivated these respondents to choose the MBA program?
RQ3: What is the level of respondent's career satisfaction, and what are the sources of it?
RQ4: What is the level of the respondent's job satisfaction after an MBA?
RQ5: In their opinion, how far they recommend future engineering graduates to pursue a career in MBA?

The next section focusses on highlighting some of the exciting contributions in management education.

LITERATURE REVIEW

An extensive literature review helped to substantiate the need for research and to identify the research gap. When engineers develop a product or service, the best one is not always best designed. They need to consider the cost, forecast the economic trends, and most importantly, the solution they provide needs to favor the financial state of the company. Engineers need to know about investment and risk management and various other aspects they undertake in the business (Kawodza, 2015). In recent years, the management education of engineers has gained importance because the engineer's role has become volatile and challenging. The technological problems that we are going to face are involving complex and challenging to solve. Being an engineer, he/she must have diverse knowledge and skills (Jablokow, 2007).

For rapid economic growth, investment in technology and training engineers is not sufficient. Thus, management education will play a crucial role, and the need was realized in the early sixties in India. To fill this gap, the Government of India has set up various management institutions such as Administrative Staff College of India (ASCI), Indian Institute of Management at Ahmedabad, Calcutta, Bangalore, and Lucknow. Later many institutions were constituted to provide management education (Mote, 1985). Over the last few years, a significant response from universities, with many new and innovative management courses and modes have been adopted. It should be acknowledged that a critical driver behind this change has been the increased emphasis on management education by an individual's interest and need in the industry.

According to the education statistics at a glance - 2018 report, 15.28% of postgraduates of the total enrolled candidates were in management. Management graduates start better in their career and get the highest paid jobs in all countries. With the advance in time, they do better over the employment life cycle. Business leaders such as Darren W Woods (CEO-ExxonMobil), Mary T Barra (CEO-General Motors), Satya Nadella (CEO-Microsoft), Sundar Pichai (CEO-Microsoft), Shantanu Narayen (CEO-Adobe Systems), D Rajkumar (CEO-Bharath Petroleum), and Rajesh Gopinath (CEO-TCS) are few names who have accomplished much after their engineering and management education.

There have been numerous studies that have studied the relation between education and success. Shelby D Hunt & Lawrence B Chonko (1987) examined the relationship between marketing education and career success in the long run. A survey of 1,000 marketing practitioners showed that marketing education does not seem to contribute to long-run success. The same study was replicated by (Bacon, 2017) applying the regression analysis on a sample of 864 marketing professionals. The results conclude that a marketing degree was positively associated with income in marketing jobs, and career satisfaction with a marketing career was not significantly associated with the major. Over the last 30 years, there has been a substantial change in the opportunities after MBA education. Van Zyl & de Villiers (2011) examined the factors influencing the career choice of accounting students in South Africa. The study considered 12 factors as a possible influence on their choice of career. The results of factor analysis indicate that the main drivers are job security, job satisfaction, aptitude for accounting, and potential future earnings. A similar study was made by Owusu, Essel-Anderson, Ossei Kwakye, Bekoe, & Ofori (2018) on students at the tertiary level in Ghana. The study considered 16 factors as career choice factors. The authors found that employability, financial prospects, and intrinsic value were the most important, and less preference was given to prestige and working conditions when considering their careers.

Most of the empirical studies have examined the influencing variables on individuals' attitudes towards careers and the importance of those factors in career choices. The literature showed that job satisfaction, social prestige, advancement opportunity, career path and flexibility, earnings potential, initial job, job security, availability of employment, nature of the job, working conditions, working hours, an element of variety and adventure, and time for a personal life that influence students' career choices (Jackling & Calero, 2006; Jackling & Keneley, 2009; Owusu et al., 2018; van Zyl & de Villiers, 2011)

Alansari (2011) examined the factors influencing career choice, job satisfaction, and public image of 117 professional librarians working in Kuwait. The study found that job satisfaction depends on the recognition of performance, performance evaluation, and job security and benefits, rewards and promotions they receive, communication about job performance, salaries, and working conditions were the leading causes of job dissatisfaction. There are many similar studies in this area and have found varying results. Gender did not show any significant association with job satisfaction (Lynch & Verdin, 1983; Mirfakhrai, 1991. In contrast, Shyu (1985) concluded that gender affected job satisfaction. Similarly, (Aldirman, 2007; Thornton, 2001) in their study found that experience affects job satisfaction.

Gajderowicz et al. (2014) examined the graduates' job satisfaction across six domains of study across six countries. The study found that, in engineering, medicine, and science area, job satisfaction is influenced by personality traits and working environment. In comparison, social science and business domains' job satisfaction is influenced by the education process. Hwang et al. (2011) studied factors that predict post MBA industry shifts among MBA graduates. The study concluded that career success factors are significant predictors of industry change. Arthur et al. (2005), Hall & Chandler (2005) researched on MBA graduate's satisfaction in terms of benefits they gain by MBA education, such as gaining knowledge, skills, and competencies significantly related to career growth and skills utilization.

The MBA program is aimed to enhance the students' knowledge in the business domain by applying practical learning methods. The evaluation anchored around student career success and satisfaction. There are a minimal number of studies in career satisfaction after the MBA program. Thus, this study is expected to contribute to this area. India has been experiencing more and more MBA aspirants every year. This, in turn, leads to a higher demand for a quality MBA program.

Therefore, the study aims at identifying the impact of management education on career satisfaction after engineering. It contributes to the existing knowledge and understanding of the field by identifying the added value of management education. Apart from the contribution to the academic literature, the findings have implications for individual graduates and business schools.

INSTRUMENT DESIGN

A structured self-administered questionnaire was considered suitable for data collection. The first section of the survey contained demographic data. The second section consisted of possible influence on their choice of career (MBA) related questions. The study considered the 16 factors based on empirical studies (Sugahara and Boland, 2009; Owusu et al., 2018). The respondents asked about their perception of the importance of 16 factors that are adopted by Owusu et al. (2018). However, the respondents were allowed to indicate any other reason that may have influenced their choice to become or not become a techno manager. The third section consisted of an academic major satisfaction scale (AMSS). AMSS was adopted from an empirical study by Nauta (2007). The questions on the range altered to the need for this study. Item no. 3 and 6 were excluded from the AMSS scale as this study focused on post MBA students. The fourth section contained questions related to career satisfaction, job satisfaction, and recommendations to others. Career satisfaction is measured using the scale developed by Greenhaus, Parasuraman, and womley (1990) and in various empirical studies were used (Seibert, Kraimer, & Crant, 2001). Job satisfaction was measured using 14 factors index scale adopted from Shelby D Hunt, Lawrence B Chonko, (1987). The respondents have given options to rate the level of importance or satisfaction they attach to these factors on a scale of 1-5 with '1' denoting not important or strongly disagree, and '5' denoting very important or strongly agree.

The sample consisted of techno managers who have completed MBA after engineering. A convenient sampling technique opted for selecting the sample. The total number of participants was 96. Data collected using an online questionnaire during April - June of the year 2020. The results computed using inferential statistics and factor analysis using IBM SPSS 26 version.

RESULTS AND DISCUSSION

Table 1 shows the results from the descriptive statistics. The characteristics are gender, MBA major, experience, income, and approximately when did they decide to do an MBA.

Table 1. Demographic characteristics of respondents

Variable	Frequency	Percent
Gender		
Male	59	61.5%
Female	37	38.5%
MBA major		
Marketing	35	36.5%
HR	21	21.9%
Finance	35	36.5%
Operations	1	1%
IT	4	4.2%
Experience		
Less than 5 years	70	73%
5 to 10 years	21	22%
More than 10 years	5	5%
Approximately when did you decide to do an MBA		
during engineering	67	69.8%
after working incorporate	13	13.5%
the decision emerged slowly over a long time	13	13.5%
I don't know	3	3.1%

Respondents have mostly completed their post-graduation in MBA after engineering. The gender distribution of males and females is 59 and 37, respectively. There was equal distribution in the specialization of marketing and finance with the highest of 35, and then 21 respondents choose HR as their major. Considering the experience, 73 percent of the respondents have less than five years, and 21per cent of the respondents have more than five and less than ten years.

The respondents asked to indicate when they have decided to pursue an MBA as their career. They have given four options to choose from the list. Table 1 shows the responses. Interestingly the majority of the respondents (69.8 percent) decided to do an MBA during engineering, and 13.5 percent of respondents opted while working. These results indicate the majority of respondents were exposed to management education and were aware of the importance of MBA in their careers. Note that only three respondents considered the MBA program without knowing. It appears that these respondents choose MBA by the option, not by choice.

The respondents asked a question related to what benefits they have received after MBA education. They were given four options initial job, salary advance, promotion, and a special bonus, and asked to choose one option. Seventy-three percent of respondents have secured the initial job, and twenty-six percent of respondents have received promotion and advancement in salary (Refer figure 1). This result indicates MBA education after engineering or engineer with management skills secures an initial job and provides fringe benefits.

Figure 1. Benefits received by respondents after MBA

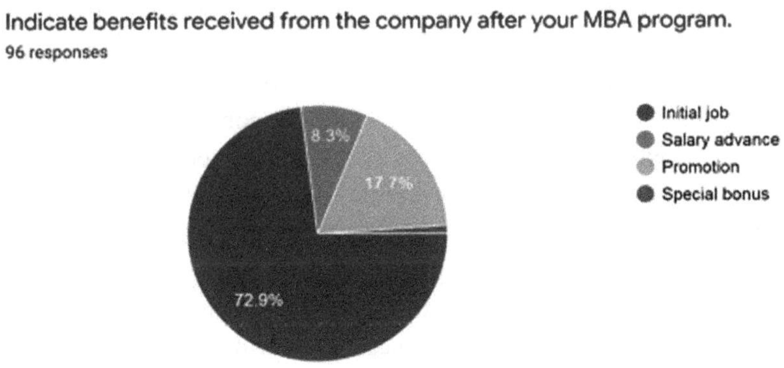

Possible Influences of Career Choice

Many factors influence individuals to choose a particular career choice. To answer what possible influencing factors motivated the respondents of this study to opt for the MBA program after engineering was further analyzed. The potential influencing factors of a career choice as an MBA were subjected to exploratory factor analysis. The principal component method of extraction criterion using the varimax rotation method used in conducting the factor analysis. The KMO test for sampling adequacy is 0.741 and indicates the data were adequate for conducting factor analysis. And the Bartlett test of sphericity shows a significant value ($p<0.01$), which suggests that the data are valid for factor analysis, as indicated in Table 2.

Table 2. KMO and Bartlett's Test

Kaiser-Meyer-Olkin Measure of Sampling Adequacy.		.741
Bartlett's Test of Sphericity	Approx. Chi-Square	493.315
	df	120
	Sig.	.000

From the 16 variables considered, five factors extracted, explaining the total variance of 63.84. Table 3 demonstrates the reduction of factors. The first factor termed as "employability and social prestige," contains indicators that characterize structured career path, advancement opportunities, job availability, the security of employment, and social prestige. The second factor named "desired working conditions" is characterized by chance to contribute, right physical working conditions, interaction with others, and sufficient time for personal life. The third factor, named "Opportunity and Job satisfaction," is characterized by an element of variety and adventure, flexibility in career options, and job satisfaction. Two items have loaded on the fourth factor and were related to the length of work hours and nature of the job, which were referred to as "Job responsibilities." The fifth factor was applied to as "earnings" had a total of two items loaded on it and characterized by initial salary and long-term earning potential. Therefore, the results from the factor analysis indicate that the possible influencing factors for career choice as an MBA after engineering can broadly be categorized into five factors, as shown in table 3.

Respondents have given an option to mention the other factors that influenced them to choose an MBA as a career choice. Interestingly, they referred to have a change in career, interest in human resource & finance specialization, and passion towards MBA.

Sources of Satisfaction

The respondents have given a list of 14 sources of job satisfaction and five sources of career satisfaction associated with the MBA as a career. The respondents have given to indicate the level of satisfaction on a five-point Likert type scale with a range of 5 for "Extremely satisfied" to 1 for "Not satisfied." Table 4 shows the data in the form of mean scores of satisfaction sources.

Table 3. Factor analysis on possible influencers

Possible Influencers	Employability and Social Prestige	Desired Working Conditions	Opportunity and Job Satisfaction	Job Responsibilities	Earnings
Structured career path	.520				
Advancement opportunities	.692				
Job availability	.764				
Security of employment	.597				
Social prestige	.715				
Chance to make a contribution		.742			
Good physical working conditions		.713			
Interaction with others		.517			
Sufficient time for personal life		.723			
An element of variety and adventure			.777		
Flexibility in career options			.606		
Job satisfaction			.729		
Length of work hours				.728	
Nature of the job				.750	
Good initial salary					.719
Good long-term earning potential					.645
Eigen values	4.711	1.941	1.475	1.070	1.024
Percentage of variance explained	29.446	12.128	9.217	6.688	6.399

Among the various sources of job satisfaction, the top five ranked sources are an opportunity to complete the work he/she start (3.80), variety in the job (3.75), opportunity to interact with others (3.72), an opportunity for independent thought and action (3.69) and freedom (3.65). These five sources of satisfaction relate to independence in doing their work and exchanging ideas with other staff. It was surprising to observe that information they receive from the superior about job performance and pay for their job are least ranked. These results indicate the organization where they are working is not maintaining transparency about their performance, and most of the respondents feel that they are underpaid. The overall mean of job satisfaction (3.27) indicates the respondents are moderately satisfied with their job (refer table 4).

Table 4. Sources of job satisfaction

Source	Mean	Rank
Satisfaction with information		
I am satisfied with the information I receive from my superior about my job performance.	3.19	13
I receive enough information from my supervisor about my job performance	3.49	9
I receive enough feedback from my supervisor about how well I am doing.	3.61	7
There is enough opportunity in my job to find out how well I am doing.	3.63	6
Satisfaction with a variety		
I am satisfied with the variety of activities my job offers.	3.40	11
I am satisfied with the freedom I have to do what I want on my job.	3.42	10
I am satisfied with the opportunities my job provides me to interact with others.	3.72	3
There is enough variety in my job.	3.75	2
I have enough freedom to do what I want in my job.	3.67	5
My job has enough opportunity for independent thought and action.	3.69	4
Satisfaction with closure		
I am satisfied with the opportunities my job gives me to complete tasks from beginning to end.	3.54	8
My job has enough opportunity to complete the work I start.	3.80	1
Satisfaction with pay		
I am satisfied with the pay I receive for my job.	2.99	14
I am satisfied with the security my job provides me.	3.39	12
Overall		
In general, I am satisfied with my job	3.27	

Table 5 represents the mean score of the sources of career satisfaction of techno managers. The top-ranked source of career satisfaction is progress in acquiring new skills (3.31) and meeting goals for advancement (3.06). These results indicate the respondents were career-oriented. It also is supported by most of the respondents choosing an MBA as a career during engineering. Therefore it shows the techno managers have predetermined goals in their careers and achieve as they progress. Finally, MBA education has been enabling them to succeed in their career. The overall mean of career satisfaction (3.83) indicates the respondents were satisfied with their career choice.

Further, to understand the relationship between gender, experience, job satisfaction, career satisfaction, and satisfaction with the major specialization opted in MBA were analyzed using the mean comparison analysis.

Table 5. Sources of career satisfaction

Source	Mean	Rank
I am satisfied with the progress I have achieved in my career	3.02	4
I am satisfied with the progress I have made toward meeting my overall career goals	3.03	3
I am satisfied with the progress I have made toward meeting my goals for income	2.82	5
I am satisfied with the progress I have made toward meeting my goals for advancement	3.06	2
I am satisfied with the progress I have made toward meeting for the development of new skills.	3.31	1
Overall		
In general, I am satisfied with my career after MBA	3.83	

The respondents have given choices to rate their level of satisfaction with the major specialization in MBA. Table 6 shows the results. The overall mean score (4.08) with the major they have opted in MBA indicates they are satisfied with the specialization they pursued (refer table 6).

Table 6. Academic major satisfaction scale

Source	Mean
I often wish I had not gotten into this major.	2.33
I wish I were happier with my choice of an academic major.	3.44
I feel good about the major I have selected.	3.92
Overall, I am happy with the major I have chosen.	4.08

To analyze the influence of gender in career satisfaction, job satisfaction, and satisfaction with the major specialization in MBA, the following hypothesis was framed and tested.

$H_0 1$: There is no gender difference in career satisfaction
$H_0 2$: There is no gender difference in job satisfaction
$H_0 3$: There is no gender difference in satisfaction with the major.

The results in table 7 show that the significance values are above the threshold value ($p > 0.05$). Therefore, the null hypothesis was accepted, and it indicates there is no gender influence on satisfaction. That means males and females are

Table 7. Mean comparison with gender

ANOVA						
		Sum of Squares	**df**	**Mean Square**	**F**	**Sig.**
Mean_AMSS	Between Groups	.200	1	.200	.477	.492
	Within Groups	39.360	94	.419		
	Total	39.560	95			
Mean_JS	Between Groups	.004	1	.004	.006	.936
	Within Groups	52.614	94	.560		
	Total	52.617	95			
Mean_CS	Between Groups	.618	1	.618	.797	.374
	Within Groups	72.902	94	.776		
	Total	73.520	95			

experiencing the same level of job satisfaction, career satisfaction, and satisfaction with an MBA major.

To analyze the role of experience in career satisfaction, job satisfaction, and satisfaction with the major specialization opted in MBA, the following hypothesis was framed and tested.

$H_0$1: There is no effect of experience in career satisfaction
$H_0$2: There is no effect of experience in job satisfaction
$H_0$3: There is no effect of experience in satisfaction with the major.

Table 8. Mean comparison with the experience

ANOVA						
		Sum of Squares	**df**	**Mean Square**	**F**	**Sig.**
Mean_AMSS	Between Groups	2.520	2	1.260	3.163	.047
	Within Groups	37.040	93	.398		
	Total	39.560	95			
Mean_JS	Between Groups	4.788	2	2.394	4.655	.012
	Within Groups	47.829	93	.514		
	Total	52.617	95			
Mean_CS	Between Groups	5.000	2	2.500	3.393	.038
	Within Groups	68.520	93	.737		
	Total	73.520	95			

The significance values in table 8 indicate that there exists a difference when compared with the experience. The significance value of the mean academic major satisfaction scale is nearly 0.05. the result shows that there is no effect of experience in satisfaction with the major. Whereas the significance values of mean job satisfaction and mean career satisfaction are below the threshold value ($p<0.05$), this indicates experience shows an effect on career satisfaction and job satisfaction.

Further, to understand which group of experiences people are having difference mean satisfaction, multiple comparisons using Scheffe test are conducted and results shown in table 9.

Table 9. Multiple comparisons with the level of experience

Dependent Variable	(I) Experience (in yrs.)	(J) Experience (in yrs.)	Mean Difference (I-J)	Std. Error	Sig.
Mean_AMSS	Less than 5 Years	5 to 10 Year s	-.325	.157	.123
		Greater Than 10 Years	.339	.292	.512
	5 to 10 Year s	Less than 5 Years	.325	.157	.123
		Greater Than 10 Years	.664	.314	.112
	Greater Than 10 Years	Less than 5 Years	-.339	.292	.512
		5 to 10 Year s	-.664	.314	.112
Mean_JS	Less than 5 Years	5 to 10 Year s	-.471*	.178	.035
		Greater Than 10 Years	.393	.331	.497
	5 to 10 Year s	Less than 5 Years	.471*	.178	.035
		Greater Than 10 Years	.865	.356	.058
	Greater Than 10 Years	Less than 5 Years	-.393	.331	.497
		5 to 10 Year s	-.865	.356	.058
Mean_CS	Less than 5 Years	5 to 10 Year s	-.466	.213	.097
		Greater Than 10 Years	.451	.397	.527
	5 to 10 Year s	Less than 5 Years	.466	.213	.097
		Greater Than 10 Years	.918	.427	.105
	Greater Than 10 Years	Less than 5 Years	-.451	.397	.527
		5 to 10 Year s	-.918	.427	.105

Scheffe test

*. The mean difference is significant at the 0.05 level.

To study the influence of satisfaction with the major in MBA on career satisfaction and job satisfaction mean comparisons made. ANOVA results shown in table 10 indicate there is an influence of satisfaction with the major on a career as well as job satisfaction. Further, posthoc tests carried to understand which level of experience affects mean career satisfaction, job satisfaction, and major specialization in MBA. The multiple comparison table indicates that there is exists an insignificant difference in mean career satisfaction and mean academic major satisfaction. The result shows the specialization in MBA does not affect career satisfaction (refer to table 10 & 11).

Table 10. Comparison of major satisfaction

ANOVA					
Mean_AMSS					
	Sum of Squares	**df**	**Mean Square**	**F**	**Sig.**
Between Groups	12.172	2	6.086	10.468	.000
Within Groups	165.697	285	.581		
Total	177.869	287			

Table 11. Multiple comparisons with major specialization

Dependent Variable: Mean_AMSS						
Scheffe						
(I) Mean_CS	**(J) Mean_CS**	**Mean Difference (I-J)**	**Std. Error**	**Sig.**	**95% Confidence Interval**	
					Lower Bound	**Upper Bound**
MEAN JS	MEAN CS	-.46935*	.11006	.000	-.7402	-.1985
	MEAN AMSS	-.39271*	.11006	.002	-.6635	-.1219
MEAN CS	MEAN JS	.46935*	.11006	.000	.1985	.7402
	MEAN AMSS	.07664	.11006	.785	-.1942	.3474
MEAN AMSS	MEAN JS	.39271*	.11006	.002	.1219	.6635
	MEAN CS	-.07664	.11006	.785	-.3474	.1942
*. The mean difference is significant at the 0.05 level.						

Recommend Other Engineers to Become Techno Managers

The respondents have given options to express their opinion about the recommendation to others to pursue a career in MBA. One question was asked "according to your career choice and satisfaction, do you recommend other engineers to pursue a career in MBA?" and were given a five-point scale ranging from 5 for "Extremely" to 1 for "Not at all." Figure 2 shows the responses of the participants. Interestingly 93 respondents mentioned they would recommend others to pursue a career in MBA based on their satisfaction.

Figure 2. Recommendation to others

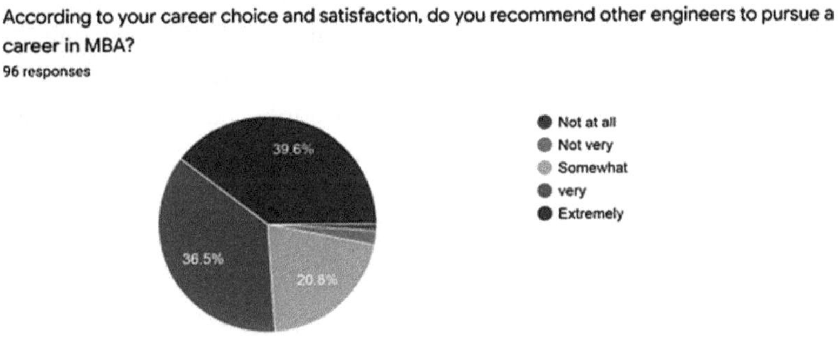

MANAGERIAL IMPLICATIONS

The study focused on identifying some of the critical findings that would help several stakeholders decide on management education after an engineering degree. As a first step, the study helps in identifying some of the essential factors associated with choosing an MBA after engineering. People who pursued an MBA have realized that there is variety in their job, and there are excellent opportunities to complete the work. Also, most of them were satisfied with the progress they made towards the development of new skills. Next, the critical factor that raised was "employability and social prestige" that contained indicators that characterized structured career path, advancement opportunities, job availability, the security of employment, and social prestige. The results show MBA education after engineering paves the way to a career path, advanced opportunities, availability, security, and social prestige. Also, the relationship between gender, experience to satisfaction examined using detailed analysis. The results indicate that gender had no influence, but experience had an impact on satisfaction. This shows that the respondent found the experience as it grows to be a more important metric for satisfaction because his career develops. He gets better opportunities while he matures up the roles.

The study found that MBA education has enhanced the value in the careers of respondents through promotions and salary advances. Further, the study found that male and female techno managers are experiencing the same level of satisfaction and experience show an effect on satisfaction. The primary sources of satisfaction are the opportunity to complete the work, variety in the job, and opportunity that the current position provides when considering job satisfaction. Whereas career satisfaction sources progress, they made in developing new skills and meeting goals for advancement. The mean score of job satisfaction (3.27) and career satisfaction (3.83) indicates the techno managers are satisfied in their career in MBA.

Management education gives them an insight into all the aspects that make up a business, such as finance, marketing, and operational management. It helps them to develop soft skills, such as communication and human resource management. The main benefits of the technical management degree are that it is tailored to suit the company's particular needs and is studied on a modular basis, allowing students to select elements that will be of direct benefit to their jobs.

CONCLUSION

The chapter addresses some of the critical questions that would be of immense help to the engineering graduates who would want to pursue their career in management education. It will also be a clear guide for engineers to understand the significance of management courses. The findings have clearly stated that there is a definite inclination towards management education. The study has also indicated that there is career satisfaction after pursuing management education for upcoming techno managers. This chapter guides and shares a message about the significance of an MBA and career after MBA. The results will be useful in informing and guiding organizations in building an employee-friendly environment and creating higher levels of job satisfaction. Thus, management can capitalize on their human capital.

REFERENCES

Alansari, H. A. (2011). Career choice, satisfaction, and perceptions about their professional image: A study of Kuwaiti librarians. *Library Review*, *60*(7), 575–587. doi:10.1108/00242531111153597

Aldirman, F. A. (2007). Job satisfaction of librarians at learning resource centers in the Kingdom of Saudi Arabia. *JAFLI (Aalam)*, *1*, 213–247.

Bacon, D. R. (2017). Revisiting the Relationship Between Marketing Education and Marketing Career Success. *Journal of Marketing Education*, *39*(2), 109–123. doi:10.1177/0273475317710061

Baruch, Y., & Peiperl, M. (2006). The Impact of an MBA on Graduate Careers. *Human Resource Management Journal*, *10*(2), 69–90. doi:10.1111/j.1748-8583.2000.tb00021.x

Ben-Haim, Y. (2000). Why the Best Engineers Should Study Humanities. *International Journal of Mechanical Engineering Education*, *28*(3), 195–200. doi:10.7227/IJMEE.28.3.2

Caldecote, V. (1986). Engineers in the 21st century. *International Journal of Technology Management*, *I*(1/2), 31–34.

Gajderowicz, T., Grotkowska, G., & Wincenciak, L. (2014). Graduates' job satisfaction across domains of study. *International Journal of Manpower*, *35*(4), 470–499. doi:10.1108/IJM-05-2013-0109

Greenhaus, J. H., Parasuraman, S., & Wormley, W. M. (1990). Effects of race on organizational experiences, job performance evaluations, and career outcomes. *Academy of Management Journal*, *33*, 64–86.

Hall, D. T., & Chandler, D. E. (2005). Psychological success: When the career is calling. *Journal of Organizational Behavior*, *26*(2), 155–176. doi:10.1002/job.301

Hwang, A., Bento, R., & Ben Arbaugh, J. (2011). Post-MBA industry shifts: An investigation of career, educational and demographic factors. *Career Development International*, *16*(6), 592–615. doi:10.1108/13620431111178344

Jablokow, K. (2007). Engineers as Problem-Solving Leaders. *IEEE Technology and Society*, *26*(4), 29–35. doi:10.1109/MTS.2007.911075

Jackling, B., & Calero, C. (2006). Influences on undergraduate students' intentions to become qualified accountants: Evidence from Australia. *Accounting Education*, *15*(4), 419–438. doi:10.1080/09639280601011115

Jackling, B., & Keneley, M. (2009). Influences on the supply of accounting graduates in Australia: A focus on international students. *Accounting and Finance*, *49*(1), 141–159. doi:10.1111/j.1467-629X.2008.00273.x

Kawodza, A. (2015). The Special Link between Finance and Engineering. In *International Conference on Mechanical and Industrial Engineering (ICMIE'15)* (pp. 20–22). Academic Press.

Lynch, B., & Verdin, J. (1983). Job satisfaction in libraries. *The Library Quarterly, 53*(4), 434–447. doi:10.1086/601406

Hunt, Chonko, & Wood. (1987). Marketing education and marketing success: Are they related? *Journal of Marketing Education, 9*(2), 183–205.

Mirfakhrai, M. K. (1991). Correlates of job satisfaction among academic librarians in the United States. *Journal of Library Administration, 14*(1), 117–131. doi:10.1300/J111v14n01_08

Mote, V. L. (1985). Development of Management Education in India. *Vikalpa, 10*(4), 371–386. doi:10.1177/0256090919850402

Nauta, M. M. (2007). Assessing college students' satisfaction with their academic majors. *Journal of Career Assessment, 15*(4), 446–462. doi:10.1177/1069072707305762

Owusu, G. M. Y., Essel-Anderson, A., Ossei Kwakye, T., Bekoe, R. A., & Ofori, C. G. (2018). Factors influencing career choice of tertiary students in Ghana: A comparison of science and business majors. *Education + Training, 60*(9), 992–1008. doi:10.1108/ET-04-2017-0050

Seibert, S. E., Kraimer, M. L., & Crant, J. M. (2001). What do proactive people do? A longitudinal model linking proactive personality and career success. *Personnel Psychology, 54*(4), 845–874. doi:10.1111/j.1744-6570.2001.tb00234.x

Shyu, J. F. (1985). A study on job satisfaction of the university library staff. *Journal of Library and Information Science, 11*, 207–234.

Sugahara, S., & Boland, G. (2009). The Accounting Profession as a Career Choice for Tertiary Business Students in Japan-A Factor Analysis. *Accounting Education, 18*(3), 255–272. doi:10.1080/09639280701820035

Thornton, J. K. (2001). African American female librarians: A study of job satisfaction. *Journal of Library Administration, 33*(1/2), 141–164. doi:10.1300/J111v33n01_10

Van Zyl, C., & de Villiers, C. (2011). Why some students choose to become chartered accountants (and others do not). *Meditari Accountancy Research, 19*(1/2), 56–74. doi:10.1108/10222521111178637

Compilation of References

ABET Engineering Accreditation Commission. (2010). *Criteria for accrediting engineering programs*. ABET.

Adair, D., Alman, S. W., Budzick, D., Grishman, L. M., Mancini, M. E., & Thackaberry, S. (2014). Many shades of MOOCs. *Internet Learning, 3*(1), 53-72. http://citeseerx.ist.psu.edu/viewdoc/download?doi=10.1.1.883.3241&rep=rep1&type=pdf

Aditomo, A., Goodyear, P., Bliuc, A.-M., & Ellis, R. A. (2013). Inquiry-based learning in higher education: Principal forms, educational objectives, and disciplinary variations. *Studies in Higher Education, 38*(9), 1239–1258. doi:10.1080/03075079.2011.616584

Alammary, A., Sheard, J., & Carbone, A. (2014). Blended learning in higher education: Three different design approaches. *Australasian Journal of Educational Technology, 30*(4), 440-454. http://doi:10.14742/ajet.693

Alamri, M. (2011). Higher Education in Saudi Arabia. *Journal of Higher Education Theory and Practice, 11*(4), 88–91.

Alansari, H. A. (2011). Career choice, satisfaction, and perceptions about their professional image: A study of Kuwaiti librarians. *Library Review, 60*(7), 575–587. doi:10.1108/00242531111153597

Aldirman, F. A. (2007). Job satisfaction of librarians at learning resource centers in the Kingdom of Saudi Arabia. *JAFLI (Aalam), 1*, 213–247.

Allen, S. J. (2007). Adult learning theory and leadership development. *Leadership Review, 7*, 26–37.

Alraimi, K. M., Zo, H., & Ciganek, A. P. (2015). Understanding the MOOCs continuance: The role of openness and reputation. *Computers & Education, 80*, 28–38. doi:10.1016/j.compedu.2014.08.006

Amagoh, F. (2009). Leadership development and leadership. *Management Decision, 47*(6), 989–999. doi:10.1108/00251740910966695

Anastasiades, P. (2002a). In Towards the global information society: the enactment of a regulatory framework as a factor of transparency and social cohesion. Lecture Notes in Computer Science, 2510, 527–535. doi:10.1007/3-540-36087-5_62

Andreasen, R. J., & Wu, C. H. (1999). Study abroad program as an experiential capstone course: A proposed model. *Journal of International Agricultural and Extension Education, 6*(2), 69–78. doi:10.5191/jiaee.1999.06209

Antonakis, J. (2001). *The validity of the transformational, transactional, and laissez-faire leadership model as measured by the Multifactor Leadership Questionnaire (MLQ 5X)* (Unpublished PhD thesis). Minneapolis, MN: Walden University ScholarWorks.

Arciszewski, T. (2006). Civil engineering crisis. *Leadership and Management in Engineering, 6*(1), 26–30. doi:10.1061/(ASCE)1532-6748(2006)6:1(26)

Athreya, K. S., & Kalkhoff, M. T. (2010). The Engineering Leadership Program: A co-curricular learning environment by and for students. *Journal of STEM Education, 11*(3-4), 70-74.

Augusti, G. (2009). EUR-ACE: the European Accreditation system of engineering education and its global context. In *Engineering Education Quality Assurance* (pp. 41–49). Springer. doi:10.1007/978-1-4419-0555-0_3

Aviv, R., Erlich, Z., Ravid, G., & Geva, A. (2003). Network analysis of knowledge construction in asynchronous learning networks. *Journal of Asynchronous Learning Networks, 7*(3), 1–23.

Avolio, B. J., & Hannah, S. T. (2008). Developmental readiness: Accelerating leader development. *Consulting Psychology Journal: Practice and Research, 60*(4), 331–347. doi:10.1037/1065-9293.60.4.331

Aydın, B., & Demirer, V. (2016). Flipping the drawbacks of flipped classroom: Effective tools and recommendations. *Journal of Educational and Instructional Studies in the World, 6*(1), 33–40. https://www.researchgate.net/publication/303813559_FLIPPING_THE_DRAWBACKS_OF_FLIPPED_CLASSROOM_EFFECTIVE_TOOLS_AND_RECCOMENDATIONS

Bacon, D. R. (2017). Revisiting the Relationship Between Marketing Education and Marketing Career Success. *Journal of Marketing Education, 39*(2), 109–123. doi:10.1177/0273475317710061

Baker, S., & Gerlowski, D. (2007). Team effectiveness and leader-follower agreement: An empirical study. *The Journal of American Academy of Business, Cambridge, 12*(1), 15–23.

Bakhtin, M., & Ghāsemipour, G. (2011). The problem of speech genres. *Literary Criticism, 4*(15), 114–136.

Balkin, J. M., & Sonnevend, J. (2016). The digital transformation of education. In C. Greenhow, J. Sonnevend., & C. Agur (Eds.), *Education and social media: Toward a digital future* (pp. 9-25). MIT Press. https://books.google.com/books?id=6jeNDwAAQBAJ&printsec=frontcover&dq

Bandura, A. (1977). *A Social Learning Theory. Prentice Hall.*

Barak, M., Watted, A., & Haick, H. (2016). Motivation to learn in massive open online courses: Examining aspects of language and social engagement. *Computers & Education, 94*, 49-60. http://doi:10.1016/j.compedu.2015.11.010

Barrows, H. S., & Tamblyn, R. M. (1980). *Problem-based learning: An approach to medical education*. Springer.

Barton, J. (1995). Conducting Effective Classroom Discussions. *Journal of Reading*, *38*(5), 346–350.

Baruch, Y., & Peiperl, M. (2006). The Impact of an MBA on Graduate Careers. *Human Resource Management Journal*, *10*(2), 69–90. doi:10.1111/j.1748-8583.2000.tb00021.x

Bass, B. M. (1990). From Transactional to Transformational Leadership: Learning to Share the Vision. *Organizational Dynamics*, *18*(3), 19–32. doi:10.1016/0090-2616(90)90061-S

Batiz-Lazo, B., & Efthymiou, L. (2016). *The Book of Payments: Historical and Contemporary Views on the Cashless Economy*. Palgrave McMillan. doi:10.1057/978-1-137-60231-2

Becker, F. S. (2006). Globalization, curricula reform and the consequences for engineers working in an international company. *European Journal of Engineering Education*, *31*(3), 261–272. doi:10.1080/03043790600644749

Benefield, L. D., Trentham, L. L., Khodadaki, K., & Walker, W. F. (1997). Quality improvement in a college engineering instructional program. *Journal of Engineering Education*, *86*(1), 57–68. doi:10.1002/j.2168-9830.1997.tb00266.x

Ben-Haim, Y. (2000). Why the Best Engineers Should Study Humanities. *International Journal of Mechanical Engineering Education*, *28*(3), 195–200. doi:10.7227/IJMEE.28.3.2

Bennett, S., Bishop, A., Dalgarno, B., Waycott, J., & Kennedy, G. (2012). Implementing web 2.0 technologies in higher education: A collective case study. Computers & *Education, 59*(2), 524–534. http://doi:10.1016/j.compedu.2011.12.022

Bennis, W., & Nanus, B. (1985). Strategies for Taking Charge. In *Leaders*. Harper.

Bergfjord, O. J., & Heggernes, T. (2016). Evaluation of a “flipped classroom” approach in management education. *Journal of University Teaching & Learning Practice*, *13*(5), 1–15. http://ro.uow.edu.au/jutlp/vol13/iss5/17

Berg, M. E., & Karlsen, J. T. (2013). Managing Stress in Projects Using Coaching Leadership Tools. *Engineering Management Journal*, *25*(4), 52–61. doi:10.1080/10429247.2013.11431995

Bhardwa, S. (2018). *The least and most popular undergraduate courses in the UK*. Available online: https://www.timeshighereducation.com/student/news/least-and-most-popular-undergraduate-courses-uk

Biggs, J. B. (2003). *Teaching for quality learning at university* (2nd ed.). Open University Press/Society for Research into Higher Education.

Binkley, M., Erstad, O., Herman, J., Raizen, S., Ripley, M., Miller-Ricci, M., & Rumble, M. (2012). Defining twenty-first century skills. In *Assessment and teaching of 21st century skills* (pp. 17–66). Springer Netherlands. doi:10.1007/978-94-007-2324-5_2

Blumenfeld, P. C., Marx, R. W., Soloway, E., & Krajcik, J. (1996). Learning with Peers: From small group cooperation to Collaborative Communities. *Educational Research, 25*(8), 37–40. doi:10.3102/0013189X025008037

Boling, E., Hough, M., Krinsky, H., Saleem, H., & Stevens, M. (2012). Cutting the distance in distance education: Perspectives on what promotes positive, online learning experiences. *The Internet and Higher Education, 15*(2), 118–126. doi:10.1016/j.iheduc.2011.11.006

Bolman, L. G., & Deal, T. E. (2017). *Reframing Organizations: Artistry, choice, and leadership* (6th ed.). John Wiley and Sons, Inc. doi:10.1002/9781119281856

Bonk, C. J., & Dennen, V. P. (1999). Teaching on Web: With a little help from my pedagogical friends. *Journal of Computing in Higher Education, 11*(1), 3–28. doi:10.1007/BF02940840

Boud, D. (1989). The Role of Self-Assessment in Student Grading. *Assessment and Evaluation in Higher Education*, (14), 20–30.

Boud, D. (1995). *Enhancing learning through self-assessment* (1st ed.). Kogan Page.

Boud, D., Cohen, R., & Sampson, J. (2014). *Peer Learning in higher Education: learning from and with each other*. Routledge. doi:10.4324/9781315042565

Boyatzis, R. E. (2009). Leadership development from a complexity perspective. *Consulting Psychology Journal: Practice and Research, 60*(4), 298–313. doi:10.1037/1065-9293.60.4.298

Boyatzis, R., Rochford, K., & Cavanagh, K. V. (2017). Emotional intelligence competencies in engineer's effectiveness and engagement. *Career Development International, 22*(1), 70–86. doi:10.1108/CDI-08-2016-0136

Bralic, A., & Divjak, B. (2018). Integrating MOOCs in traditionally taught courses: Achieving learning outcomes with blended learning. *International Journal of Educational Technology in Higher Education, 15*(1), 1-16. http://doi:10.118641239-017-0085-7

Brame, C. (2019, November 14). *Flipping the Classroom.* Retrieved January 30, 2020, from https://cft.vanderbilt.edu/guides-sub-pages/flipping-the-classroom/

Branch, R. M. (2009). *Instructional design: The ADDIE approach*. Springer Science & Business Media. doi:10.1007/978-0-387-09506-6

Brennan, J., Ryan, S., Ranga, M., Broek, S., Durazzzi, N., & Kamphuis, B. (2014, February 2014). *Study on innovation in higher education.* http://eprints.lse.ac.uk/55819/

Brown, M., Millichap, N., & Dehoney, J. (2015, April 27). *The Next Generation Digital Learning Environment: A Report on Research*. Retrieved from https://library.educause.edu/resources/2015/4/the-next-generation-digital-learning-environment-a-report-on-research

Brungardt, C., Greenleaf, J., Brungardt, C., & Arensdorf, J. (2006). Majoring in leadership: A review of undergraduate leadership degree programs. *Journal of Leadership Education, 5*(1), 4–25. doi:10.12806/V5/I1/RF1

Bryant, P. (2018). Making education better: Implementing pedagogical change through technology in a modern institution. In A. Zorn, J. Haywood, & J. Glachant (Eds.), *Higher education in the digital age: Moving academia online* (pp. 35-50). Edward Elgar Publishing. https://www.e-elgar.com/shop/higher-education-in-the-digital-age

Bryman, A. (2007). Effective leadership in higher education: A literature review. *Studies in Higher Education, 32*(6), 693–710. doi:10.1080/03075070701685114

Buchanan, S., Harlan, M. A., Bruce, C. S., & Edwards, S. L. (2016). Inquiry based learning models, information literacy, and student engagement: A literature review. *School Libraries Worldwide, 22*(2), 23–39.

Burke, P. F., Schuck, S., Aubusson, P., Kearney, M., & Frischknecht, B. (2017). Exploring teacher pedagogy, stages of concern and accessibility as determinants of technology adoption. *Technology, Pedagogy, and Education, 27*(2), 149–163. http://doi:10.1080/1475939x.2017.1387602

Butler, C. J., & Chinowsky, P. S. (2006). Emotional intelligence and leadership behavior in construction executives. *Journal of Management Engineering, 22*(3), 119–125. doi:10.1061/(ASCE)0742-597X(2006)22:3(119)

Caldecote, V. (1986). Engineers in the 21st century. *International Journal of Technology Management, 1*(1/2), 31–34.

Callanan, G. A. (2004). What would Machiavelli think? An overview of the leadership challenges in team-based structures. *Team Performance Management, 10*(3/4), 77–83. doi:10.1108/13527590410545072

Carter, D. F., Ro, H. K., Alcott, B., & Lattuca, L. R. (2016). Co-Curricular Connections: The Role of Undergraduate Research Experiencess in Promoting Engineering Students' Communication, Teamwork, and Leadership Skills. *Research in Higher Education, 57*(3), 363–393. doi:10.100711162-015-9386-7

Cartwright, S., & Gale, A. G. (1995). Project management: Different gender, different culture? *Leadership and Organization Development Journal, 16*(4), 12–16. doi:10.1108/01437739510089058

Caruso, D. R., & Salovey, P. (2004). *The Emotionally Intelligent Manager: How to Develop and Use the Four Key Emotional Skills of Leadership*. Jossey-Bass.

Cheng, K. H., Liang, J. C., & Tsai, C. C. (2015). Examining the role of feedback messages in undergraduate students' writing performance during an online peer assessment activity. *The Internet and Higher Education, 25*(4), 78–84. doi:10.1016/j.iheduc.2015.02.001

Cheng, X., Fu, S., Han, Y., & Zarifis, A. (2017). Investigating the Individual Trust and School Performance in Semi-Virtual Collaboration Groups. *Information Technology & People, 30*(3), 691–707. doi:10.1108/ITP-01-2016-0024

Cheng, X., Su, L., & Zarifis, A. (2019). Designing a talents training model for cross-border e-commerce: A mixed approach of problem-based learning with social media. *Electronic Commerce Research, 19*(4), 801–822. doi:10.100710660-019-09341-y

Cheng, X., Wang, X., Huang, J., & Zarifis, A. (2016). An Experimental Study of Satisfaction Response: Evaluation of Online Collaborative Learning. *International Review of Research in Open and Distributed Learning, 17*(1), 60–78. doi:10.19173/irrodl.v17i1.2110

Chin, R. J. (2015). Examining teamwork and leadership in the fields of public administration, leadership, and management. *Team Performance Management, 21*(3/4), 199–216. doi:10.1108/TPM-07-2014-0037

Chiu, T. K. F., & Hew, K. F. (2018). Factors influencing peer learning and performance in MOOC asynchronous online discussion forum. *Australasian Journal of Educational Technology, 34*(4). Advance online publication. doi:10.14742/ajet.3240

Chiu, T. K. F., & Mok, I. A. C. (2017). Learner expertise and mathematics different order thinking skills in multimedia learning. *Computers & Education, 107*, 147–164. doi:10.1016/j.compedu.2017.01.008

Cho, K., & MacArthur, C. (2010). Student revision with peer and expert reviewing. *Learning and Instruction, 20*(4), 328–338. doi:10.1016/j.learninstruc.2009.08.006

Ciulla, J. B., Day, D. D., & Antonakis, J. (2011). *Ethics Effectiveness: The Nature of Good Leadership*. Sage Publications, Inc.

Clark, R. C., & Mayer, R. E. (2007). *E-Learning and the Science of Instruction: Proven Guidelines for Consumers and Designers of Multimedia Learning*. Pfeiffer.

Cogill, J. (2010). A model of pedagogical change for the evaluation of interactive whiteboard practice. In M. Thomas & E. C. Schmid (Eds.), *Interactive whiteboards for education: Theory, research, and practice* (pp. 162–178). Information Science Reference. http://doi:10.4018/978-1-61520-715-2.ch011

Conole, G. (2016). MOOCs as disruptive technologies: Strategies for enhancing the learner experience and quality of MOOCs. *Revista de Educación a Distancia, 50*(2), 1-7. http://doi:10.6018/red/50/2

Conrad, D. (2006). E-Learning and social change: An apparent contradiction. In M. Beaudoin (Ed.), Perspectives on higher education in the digital age, (pp. 21–33). New York: Nova Science Publishers.

Core Curriculum and Undergraduate Services Committee. (2003, June). *Proposal for revising the general education component of engineering undergraduate curricula*. Retrieved November 12, 2003, from www.eng.ohio-state.edu/faculty/forms/Eng_GEC_Proposal.pdf

Cornelissen, J. (2017). Editor's comments: Developing propositions, a process model, or a typology? Addressing the challenges of writing theory without a boilerplate. *Academy of Management Review, 42*(1), 1–9. doi:10.5465/amr.2016.0196

Costello, E., Holland, J., & Kirwan, C. (2018). The future of online testing and assessment: Question quality in MOOCs. *International Journal of Educational Technology in Higher Education, 15*(1), 5-10. http://doi:10.118641239-018-0124-z

Cotič, M., & Zuljan, M. V. (2009). Problem-based instruction in mathematics and its impact on the cognitive results of the students and on affective-motivational aspects. *Educational Studies*, *35*(3), 297–310. doi:10.1080/03055690802648085

Cox, M. F., Berry, C. A., & Smith, K. A. (2009). Development of a Leadership, Policy, and Change Course for Science, Technology, Engineering, and Mathematics Graduate Students. *Journal of STEM Education, 10*(3-4), 9-16.

Cox, M. F., Cekic, O., & Adams, S. G. (2010). Developing Leadership Skills of Undergraduate Engineering Students: Perspectives from engineering faculty. *Journal of STEM Education, 11*(3-4), 22-33.

Cox, M. F., Cekic, O., Ahn, B., & Zhu, J. (2012). Engineering professionals' expectations of undergraduate engineering students. *Leadership and Management in Engineering*, *12*(2), 60–70. doi:10.1061/(ASCE)LM.1943-5630.0000173

D'Cruz, H., Gillingham, P., & Melendez, S. (2007). Reflexivity, its meanings and relevance for social work: A critical review of the literature. *British Journal of Social Work*, *37*(1), 73–90. doi:10.1093/bjsw/bcl001

Daniel, S. J., Vazquez Cano, E., & Gisbert, M. (2015). The future of MOOCs: Adaptive learning or business model? *Universities and Knowledge Society Journal, 12*(1), 64-73. http://doi:10.7238/rusc.v12i1.2475

Daspit, J., Tillman, C. J., Boyd, N. G., & Mckee, V. (2013). Cross-functional team effectiveness: An examination of internal team environment, shared leadership and cohesion influences. *Team Performance Management*, *19*(1/2), 32–56. doi:10.1108/13527591311312088

De Laat, M., Lally, V., Lipponen, L., & Simons, R.-J. (2006). Analysing student engagement with learning and tutoring activities in networked learning communities: A multi-method approach. *International Journal of Web Based Communities*, *2*(4), 394–412. doi:10.1504/IJWBC.2006.011766

Delbridge, R., & Fiss, P. C. (2013). Editors' comments: Styles of theorizing and the social organization of knowledge. *Academy of Management Review*, *38*(3), 325–331. doi:10.5465/amr.2013.0085

Delors, J. (1997). Nell'educazione un Tesoro. Roma: Armando.

Denis, J.-L., Langley, A., & Sergi, V. (2012). Leadership in the Pluran. *The Academy of Management Annals*, *6*(1), 211–283. doi:10.5465/19416520.2012.667612

Desai, M., Hart, J., & Richards, T. (2009). E-learning: Paradigm shift in education. *Education*, *129*(2), 327–334.

Design-Based Research Collective. (2003). Design-Based Research: An emerging paradigm for educational inquiry. *Educational Researcher*, *32*(1), 5–8. doi:10.3102/0013189X032001005

Dewey, J. (1938). *Experience and education.* Collier Macmillan Publishers.

Dillenbourg, P. (2002). Over-scripting CSCL: The risks of blending collaborative learning with instructional design. In P.A. Kirschner (Ed.), Three worlds of CSCL. Can we support CSCL (pp. 61-91). Heerlen: Open Universiteit Nederland

Dillenbourg, P. (1999). What do you mean by collaborative learning? In P. Dillenbourg (Ed.), *Collaborative-learning: Cognitive and Computational Approaches* (pp. 1–19). Elsevier.

DiSessa, A., & Minstrell, J. (1998). Cultivating conceptual change with benchmark lessons. In J. Green & S. Goldman (Eds.), *Thinking Practices in Mathematics and Science Learning. Mahwah.* Lawrence Erlbaum Associates.

Domagalski, T. A. (1999). Emotion in organizations: Main currents. *Human Relations*, *52*(6), 833–852. doi:10.1177/001872679905200607

Driskell, J. E., Goodwin, G. F., O'Shea, P. G., & Salas, E. (2006). What Makes a Good Team Player? Personality and Team Effectiveness. *Group Dynamics*, *10*(4), 249–271. doi:10.1037/1089-2699.10.4.249

Duan, C., & Hill, C. E. (1996). The current state of empathy research. *Journal of Counseling Psychology*, *43*(3), 261–274. doi:10.1037/0022-0167.43.3.261

Dugan, J. P., Fath, K. Q., Howes, S. D., Lavelle, K. R., & Polanin, J. R. (2013). Developing the Leadership Capacity and Leader Efficacy of College Women in Science, Technology and Math fields. *The Journal of Leadership Studies*, *7*(3), 6–23. doi:10.1002/jls.21292

Dulaimi, M. F. (2005). The influence of academic education and formal training on the project manager's behavior. *Journal of Construction Research*, *6*(01), 179–193. doi:10.1142/S1609945105000328

Dulaimi, M. F., & Langford, D. (1999). Job behaviour of construction project managers: Determinants and assessment. *Journal of Construction Engineering and Management*, *125*(4), 256–264. doi:10.1061/(ASCE)0733-9364(1999)125:4(256)

Edström, K., & Kolmos, A. (2014). PBL and CDIO: Complementary models for engineering education development. *European Journal of Engineering Education*, *39*(5), 539–555. doi:10.1080/03043797.2014.895703

Educationaldatamining.org. (n.d.). Retrieved from http://educationaldatamining.org/

Efthymiou, L., Orphanidou, Y., & Panayiotou, G. (2019). The Latest from the Tourism Front: Technology, Innovation and Disruption. *The European Financial Review, 4*(5), 39-43. Online at: https://www.europeanfinancialreview.com/the-latest-from-the-tourism-front-technology-innovation-and-disruption/

Efthymiou, L. (2018). Worker body-art in upper-market hotels: Neither accepted, nor prohibited. *International Journal of Hospitality Management*, *74*, 99–108. doi:10.1016/j.ijhm.2018.02.012

ENAEE. (n.d.). Retrieved January 30, 2020, from https://www.enaee.eu/

Engeström, Y., & Escalante, V. (1996). Mundane tool or object of affection? The rise and fall of the Postal Buddy. In B. A. Nardi (Ed.), *Context and consciousness: Activity theory and human-computer interaction* (pp. 119–161). The MIT Press.

Epp, C. D., Phirangee, K., & Hewitt, J. (2017). Student Actions and Community in Online Courses: The Roles Played By Course Length and Facilitation Method. *Online Learning*, *21*(4), 53–77.

Esen, M., Bellibas, M. S., & Gumus, S. (2018). The Evolution of Leadership Research in Higher Education for Two Decades (1995-2014): A Bibliometric and Content Analysis. *International Journal of Leadership in Education*, *23*(3), 259–273. doi:10.1080/13603124.2018.1508753

European Commission. (2010). *Communication from the commission. Europe 2020: A strategy for smart, sustainable, and inclusive growth.* http://ec.europa.eu/eu2020/pdf/COMPLET%20EN%20BARROSO%20%20%20007%20-%20Europe%202020%20-%20EN%20version.pdf

Evans, D. L., Beakley, G. C., Crouch, P. E., & Yamaguchi, G. T. (1993). Attributes of engineering graduates and their impact on curriculum design. *Journal of Engineering Education*, *82*(4), 203–211. doi:10.1002/j.2168-9830.1993.tb01075.x

Farr, J. V., & Brazil, D. M. (2009). Leadership skills development for engineers. *Engineering Management Journal*, *21*(1), 3–8. doi:10.1080/10429247.2009.11431792

Ferguson, R., & Sharples, M. (2014). Innovative pedagogy at massive scale: Teaching and learning in MOOCs. *Lecture Notes in Computer Science, 2014*(2014), 98–111. http://doi:10.1007/978-3-319-11200-8_8

Fitzgerald, M. (2003). Corporate EQ: Creating an emotionally-intelligent workplace. *HR Professional*, *20*(1), 46–49.

Freedman, A., & Artemeva, N. (1998). Learning to teach writing to engineers. *Canadian Journal for Studies in Discourse and Writing/Rédactologie, 14*(1).

Freitas, A., & Paredes, J. (2018). Understanding the faculty perspectives influencing their innovative practices in MOOCs/SPOCs: A case study. *International Journal of Educational Technology in Higher Education, 15*(1), 5-10. http://doi:10.118641239-017-0086-6

Gajderowicz, T., Grotkowska, G., & Wincenciak, L. (2014). Graduates' job satisfaction across domains of study. *International Journal of Manpower*, *35*(4), 470–499. doi:10.1108/IJM-05-2013-0109

Garcia, E. B. J., Tenorio, S. G. C., & Ramirez, M. S. (2015). Self-motivation challenges for student involvement in the open educational movement with MOOC. *Universities and Knowledge Society Journal, 12*(1), 91. http://doi:10.7238/rusc.v12i1.2185

George-Walker, L. D., & Keeffe, M. (2010). Self-determined blended learning: A case study of blended learning design. *Higher Education Research & Development, 29*(1), 1–13. http://doi:10.1080/07294360903277380

Gerhana, M. T. C., Mardiyana, M., & Pramudya, I. (2017), The Effectiveness of Project-Based Learning in Trigonometry. *IOP Conf. Series: Journal of Physics: Conf. Series 895*. 10.1088/1742-6596/895/1/012027

Gerson, S. J., & Gerson, S. M. (2012). *Technical communication: Process and product*. Prentice Hall.

Ghannadpour, S., Jokar, M., & Makui, A. (2018). Fuzzy analytical network process logic for performance measurement system of e-learning centers of universities. *Journal of Industrial and Systems Engineering, 11*(3), 261–280.

Gielen, S., Peeters, E., Dochy, F., Onghena, P., & Struyven, K. (2010). Improving the effectiveness of peer feedback for learning. *Learning and Instruction, 20*(4), 304–315. doi:10.1016/j.learninstruc.2009.08.007

Gillespie, F. (1998). Instructional design for the new technologies. In K. H. Gillespie (Ed.), *The Impact of Technology on Faculty Development, Life, and Work. New Directions for Teaching and Learning, 76*. Jossey-Bass. doi:10.1002/tl.7603

Gill, J., Sharp, R., Mills, J., & Franzway, S. (2008). I still wanna be an engineer! Women, education and the engineering profession. *European Journal of Engineering Education, 33*(4), 391–402. doi:10.1080/03043790802253459

Gilson, L. L., & Goldberg, C. B. (2015). Editors' comment: So, what is a conceptual paper? *Group & Organization Management, 40*(2), 127–130. doi:10.1177/1059601115576425

Gokhale, A. A. (Fall 1995). *Collaborative Learning Enhances Critical Thinking Journal of Technology Education, 7(1)* (M. Sanders, Ed.).

Goldstein, A. P., & Sorcher, M. (1974). *Changing supervisor behavior*. Pergamon.

Goleman, D. (1995). *Emotional intelligence*. Bloomsbury.

Goleman, D. (1998). *Working with Emotional Intelligence*. Bloomsbury Publishing.

Gordon, J. (2002). A perspective on team building. *The Journal of American Academy of Business, Cambridge, 2*(1), 185–188.

Grammatikou, M., Pantazatos, D., Barth, I., Corleto, A., Sansone, N., & Zimniewicz, M. (2019). Pilots Results: Usage And Lesson Learnt Within The First Pilots In Europe. *EDULEARN19 Proceedings*. doi: 10.21125/edulearn.2019.1481

Greenhaus, J. H., Parasuraman, S., & Wormley, W. M. (1990). Effects of race on organizational experiences, job performance evaluations, and career outcomes. *Academy of Management Journal, 33*, 64–86.

Grosjean, G., & Sork, T. J. (2007). Going online: Uploading learning to the virtual classroom. *New Directions for Adult and Continuing Education, 113*(113), 13–24. doi:10.1002/ace.243

Guilar, J., & Loring, A. (2008). Dialogue and community in online learning: Lessons from Royal Roads University. *Journal of Distance Education*, *22*(3), 19–40.

Guilford, J. P. (1936). *Psychometric methods*. McGraw-Hill.

Gustafson, K., & Branch, R. (2002). *Survey of Instructional Development Models* (4th ed.). Eric Publications.

Hackman, J. R., & Wageman, R. (2005). A Theory of Team Coaching. *Academy of Management Review*, *30*(2), 269–287. doi:10.5465/amr.2005.16387885

Hall, D. T., & Chandler, D. E. (2005). Psychological success: When the career is calling. *Journal of Organizational Behavior*, *26*(2), 155–176. doi:10.1002/job.301

Hall, J., Chamblee, G., & Slough, S. (2013). An examination of interactive whiteboard perceptions using the concerns-based adoption model stages of concern and the apple classrooms of tomorrow model of instructional evolution. *Journal of Technology and Teacher Education*, *21*, 301–320. https://eric.ed.gov/?id=EJ1016990

Hannah, S. T., Avolio, B., Luthans, F., & Harms, P. D. (2008). Leadership efficacy: Review and future directions. *The Leadership Quarterly*, *19*(6), 669–692. doi:10.1016/j.leaqua.2008.09.007

Hanrahan, S. J., & Isaacs, G. (2001). Assessing self- and peer-assessment: The students' views. *Higher Education Research & Development*, *20*(1), 53–69. doi:10.1080/07294360123776

Harrison, R. (2009). *Learning and development* (5th ed.). Chartered Institute of Personnel and Development.

Harvard Business Press. (2009). Measuring Performance: Expert Solutions to Everyday Challenges. *Harvard Business School Press Pocket Mentor*. Online at: https://store.hbr.org/product/measuring-performance/13280

Harvard Business Review. (2018). The Best Performing CEOs in the World. *Harvard Business Review*. https://hbr.org/2018/11/the-best-performing-ceos-in-the-world-2018

Harvey, M., & Riggio, R. E. (2011). *Leadership studies: The dialogue of disciplines*. Edward Elgar Publishing. doi:10.4337/9780857936486

Haynes, N. S., & Love, P. E. (2004). Psychological adjustment and coping among construction project managers. *Construction Management and Economics*, *22*(2), 129–140. doi:10.1080/0144619042000201330

Heijstra, T. M., & Siguroardottir, M. S. (2017). The flipped classroom: Does viewing the recordings matter? *Active Learning in Higher Education, 2017*(2017), 1-13. http://doi:10.1177/1469787417723217

Henderson, M., Selwyn, N., Finger, G., & Aston, R. (2015). Students' everyday engagement with digital technology in university: Exploring patterns of use and 'usefulness'. *Journal of Higher Education Policy and Management, 37*(3), 308-319. http://doi:10.1080/1360080X.2015.1034424

Hendry, G. D., Frommer, M., & Walker, R. A. (1999). Constructivism and Problem-based Learning. *Journal of Further and Higher Education*, *23*(3), 369–371. doi:10.1080/0309877990230306

Hernández-García, A., Acquila-Natale, E., & Chaparro-Peláez, J. (2017). Current state of Learning Management Systems' log data-based learning analytics, *17th Biennial EARLI Conference for Research on Learning and Instruction (EARLI 2017).*

Heuchan, J., & Archibald, J. (2003). *Industry and higher education: Meeting the needs of the mining engineering sector*. Retrieved November 27, 2003, from http://mine.queensu.ca/people/faculty/Archibald/Surv'03Paper.pdf

Hill, C., Corbett, C., & Rose, A. S. (2010). *Why so few? Women in science, technology, engineering, and mathematics*. AAUW.

Hoover, J. D., Giambatista, R. C., Sorenson, R., & Bommer, W. H. (2010). Assessing the Effectiveness of Whole Person Learning Pedagogy in Skill Acquisition. *Academy of Management Learning & Education*, *9*(2), 192–203.

Hornyak, M. J., & Page, D. (2004). Experiential learning: Introducing faculty and staff to a university leadership development program. *Simulation & Gaming*, *35*(4), 461–475. doi:10.1177/1046878104263974

Horton, W. (2006). *E-Learning by Design*. Pfeiffer.

Hoyt, C. L., & Blascovich, J. (2007). Leadership efficacy and Women Leaders' Responses to Stereotype Activation. *Group Processes & Intergroup Relations*, *10*(4), 595–616. doi:10.1177/1368430207084718

Huang, M. H., & Rust, R. T. (2018). Artificial intelligence in service. *Journal of Service Research*, *21*(2), 155–172. doi:10.1177/1094670517752459

Hulon, S. (2013). Face-to-face discussions versus online threaded discussions: Can we have the best of both worlds? In R. McBride & M. Searson (Eds.), *Proceedings of Society for Information Technology & Teacher Education International Conference 2013*, (pp. 577-582). Chesapeake, VA: Association for the Advancement of Computing in Education.

Hunt, Chonko, & Wood. (1987). Marketing education and marketing success: Are they related? *Journal of Marketing Education*, *9*(2), 183–205.

Hwang, A., Bento, R., & Ben Arbaugh, J. (2011). Post-MBA industry shifts: An investigation of career, educational and demographic factors. *Career Development International*, *16*(6), 592–615. doi:10.1108/13620431111178344

IATA. (2018). *Partner-taught and Self-study*. Online at: https://www.iata.org/training/delivery/Pages/distance-learning.aspx

Institute, A. P. (2016). *Implementing a Performance Measurement Framework at Essex Police*. Online at: https://www.bernardmarr.com/default.asp?contentID=1060

Jaakkola, E. (2020). *Designing conceptual articles: four approaches*. AMS Rev.

Jablokow, K. (2007). Engineers as Problem-Solving Leaders. *IEEE Technology and Society*, *26*(4), 29–35. doi:10.1109/MTS.2007.911075

Jackling, B., & Calero, C. (2006). Influences on undergraduate students' intentions to become qualified accountants: Evidence from Australia. *Accounting Education*, *15*(4), 419–438. doi:10.1080/09639280601011115

Jackling, B., & Keneley, M. (2009). Influences on the supply of accounting graduates in Australia: A focus on international students. *Accounting and Finance*, *49*(1), 141–159. doi:10.1111/j.1467-629X.2008.00273.x

Jenkins, D. M., & Owen, J. E. (2016). Who teaches leadership? A comparative analysis of faculty and student affairs leadership educators and implications for leadership learning. *Journal of Leadership Education*, *15*(2), 98–113. doi:10.12806/V15/I2/R1

Jeong, A. (2003). The Sequential Analysis of Group Interaction and Critical Thinking in Online Threaded Discussions. *American Journal of Distance Education*, *17*(1), 25–43. doi:10.1207/S15389286AJDE1701_3

Johansen, R., & Johansen, B. (2007). *Get there early: Sensing the future to compete in the present*. Berrett-Koehler Publishers.

Jonassen, D. H. (2006). *Modeling with technology: Mindtools for conceptual change*. Pearson Merrill Prentice Hall.

Jonassen, D. H. (Ed.). (2004). *Handbook of Research for Educational Communications and Technology* (2nd ed.). Lawrence Erlbaum Associates.

Joplin, L. (1981). On defining experiential education. *Journal of Experiential Education*, *4*(1), 16–20. doi:10.1177/105382598100400104

Judge, W. (2005). Adventures in creating an outdoor leadership challenge course for an EMBA program. *Journal of Management Education*, *29*(2), 284–300. doi:10.1177/1052562904265251

Kalyani, D., & Rajasekaran, K. (2018). Innovative teaching and learning. *Journal of Applied and Advanced Research, 3*(1), 23-25. http://doi:10.21839/jaar.2018.v3S1.162

Kanwar, A., & Uvalic-Trumbic, S. (2015). *A basic guide to open educational resources. Vancouver: Commonwealth learning*. http://oasis.col.org/bitstream/handle/11599/36/2015_UNESCO_COL_A-Basic-Guide-to-OER.pdf?sequence=6&isAllowed=y

Kark, R., & Eagly, A. H. (2010). Gender and leadership: Negotiating the labyrinth. In J. C. Chrisler & D. R. McCreary (Eds.), *In Handbook of gender research in psychology* (pp. 443–468). Springer. doi:10.1007/978-1-4419-1467-5_19

Kawodza, A. (2015). The Special Link between Finance and Engineering. In *International Conference on Mechanical and Industrial Engineering (ICMIE'15)* (pp. 20–22). Academic Press.

Keegan, D. (1996). *Foundations of distance education* (3rd ed.). Routledge.

Kintu, M. J., Zhu, C., & Kagambe, E. (2017). Blended learning effectiveness: The relationship between student characteristics, design features, and outcomes. *International Journal of Educational Technology in Higher Education, 14*(1), 7-14. http://doi:10.118641239-017-0043-4

Kirkwood, A., & Price, L. (2013). Technology-enhanced learning and teaching in higher education: what is "enhanced" and how do we know? A critical literature review. *Learning, Media and Technology, 39*(1), 6–36. http://doi:10.1080/17439884.2013.770404

Kirschner, P. A., Sweller, J., & Clark, R. E. (2006). Why minimal guidance during instruction does not work: An analysis of the failure of constructivist, discovery, problem-based, experiential, and inquiry-based teaching. *Educational Psychologist, 41*(2), 75–86. doi:10.120715326985ep4102_1

Klegeris, A., & Hurren, H. (2011). Impact of problem-based learning in a large classroom setting: Student perception and problem-solving skills. *Advances in Physiology Education, 35*(4), 408–415. doi:10.1152/advan.00046.2011 PMID:22139779

Knott, C., & Scragg, T. (2008). *Reflective practice in social work*. Learning Matters.

Koh, G. C.-H., Khoo, H. E., Wong, M. L., & Koh, D. (2008). The effects of problem-based learning during medical school on physician competency: A systematic review. *Canadian Medical Association Journal, 178*(1), 34–41. doi:10.1503/cmaj.070565 PMID:18166729

Kolb, D. (1984). *Experiential Learning*. Prentice Hall.

Kombarakaran, F. A., Yang, J. A., Fernandes, P. B., & Baker, M. N. (2008). EXECUTIVE COACHING: IT WORKS! *Consulting Psychology Journal: Practice and Research, 60*(1), 78–90. doi:10.1037/1065-9293.60.1.78

Komives, S., & Dugan, J. (2010). Contemporary leadership theories. In R. Cuoto (Ed.), *Political and civic leadership: A reference handbook* (pp. 111–120). Sage. doi:10.4135/9781412979337.n14

Kong & Pearson. (2002). 2, cited in Boling, E., Hough, M., Krinsky, H., Saleem, H. & Stevens, M. (2012). Cutting the distance in distance education: Perspectives on what promotes positive, online learning experiences. *The Internet and Higher Education, 15*(2), 118–126.

Kop, R., Fournier, H., & Mak, J. (2011). A pedagogy of abundance or a pedagogy to support human beings? Participant support on massive open online courses. *International Review of Research in Open and Distance Learning, 12*(7), 74–93. doi:10.19173/irrodl.v12i7.1041

Korn Ferry. (2016). *Tomorrow's engineering workforce.* Academic Press.

Kovarova, P., & Simkova, G. (2014). Evidence-based learning approach in evaluation of information literacy education. In *European Conference on Information Literacy* (vol. 492, pp. 560-569). Springer. http://doi:10.1007/978-3-319-14136-7_59

Kreber, C. (2001). Learning experientially through case studies? A conceptual analysis. *Teaching in Higher Education, 6*(2), 217–228. doi:10.1080/13562510120045203

Ktoridou, D., & Doukanari, E. (2015, November). Student generated content in higher education technology-related blogs. In *2015 international conference on interactive Mobile communication technologies and learning (IMCL)* (pp. 289-295). IEEE.

Kumar, S., & Hsiao, J. K. (2007). Engineers Learn "Soft Skills the Hard Way": Planting a Seed of Leadership in Engineering Classes. *Leadership and Management in Engineering*, *7*(1), 18–23. doi:10.1061/(ASCE)1532-6748(2007)7:1(18)

Lang, J. D., Cruse, S., McVey, F. D., & McMasters, J. (1999). Industry expectations of new engineers: A survey to assist curriculum designers. *Journal of Engineering Education*, *88*(1), 43–51. doi:10.1002/j.2168-9830.1999.tb00410.x

Langley, A. (2000). *New essays concerning human understanding*. Macmillan & Co. Ltd.

Larson, B. E., & Keiper, T. A. (2016). *Classroom Discussion and Threaded Electronic Discussion: Learning In Two Arenas*. Online at: https://www.citejournal.org/volume-2/issue-1-02/general/classroom-discussion-and-threaded-electronic-discussion-learning-in-two-arenas/

Lasakova, A., Bajzikova, L., & Dedze, I. (2017). Barriers and drivers of innovation in higher education: Case study-based evidence across ten European universities. *International Journal of Educational Development, 55*, 69–79. http://doi:10.1016/j.ijedudev.2017.06.002

Latif, D. A. (2004). Using emotional intelligence in the planning and implementation of a management skills course. *Pharmacy Education*, *4*(2), 81–89. doi:10.1080/15602210410001701685

Lave, J., & Wenger, E. (1991). *Situated learning: legitimate peripheral participation*. Cambridge University Press. doi:10.1017/CBO9780511815355

Le Boterf, G. (1994). De la compétence: Essai sur un attracteur étrange. Les Ed. de l'Organisation.

Leal Filho, W., Shiel, C., & Paco, A. (2016). Implementing and operationalising integrative approaches to sustainability in higher education: The role of project-oriented learning. *Journal of Cleaner Production*, *133*, 126–135. doi:10.1016/j.jclepro.2016.05.079

Leontyev, A., & Baranov, D. (2013). Massive open online courses in chemistry: A comparative overview of platforms and features. *Journal of Chemical Education*, *90*(11), 1533–1539. doi:10.1021/ed400283x

Lesser, E., & Storck, J. (2001). Communities of Practice and Organizational Performance. February 2001. *IBM Systems Journal*, *40*(4), 831–841. doi:10.1147j.404.0831

Leung, M.-y., Chan, Y.-S., & Olomolaiye, P. (2008). Impact of Stress on the Performance of Construction Project Managers. *Journal of Construction Engineering and Management*, *13*(8), 644–652. doi:10.1061/(ASCE)0733-9364(2008)134:8(644)

Ligorio, M. B., Andriessen, J., Baker, M., Knoller, N., & Tateo, L. (2009). *Talking over the computer. Pedagogical scenarios to blend computer and face to face interaction*. Scriptaweb.

Lindebaum, D., & Fielden, S. (2011). 'It's good to be angry': Enacting anger in construction project management to achieve perceived leader effectiveness. *Human Relations*, *64*(3), 437–458. doi:10.1177/0018726710381149

Lin, J., Cantoni, L., & Murphy, J. (2018). MOOCs in tourism and hospitality: A review. *Journal of Teaching in Travel & Tourism*, *18*(3), 217–235. doi:10.1080/15313220.2018.1428505

Lockyer, L., Heathcote, E., & Dawson, S. (2013). Informing Pedagogical Action. *The American Behavioral Scientist*, *57*(10), 1439–1459. doi:10.1177/0002764213479367

Loosemore, M., & Galea, N. (2008). Genderlect and conflict in the Australian construction industry. *Construction Management and Economics*, *26*(2), 125–135. doi:10.1080/01446190701798810

Lopes, P. N., Salovey, P., Côté, S., & Beers, M. (2005). Emotion regulation abilities and the quality of social interaction. *Emotion (Washington, D.C.)*, *5*(1), 113–118. doi:10.1037/1528-3542.5.1.113 PMID:15755224

Lowenthal, P., & Hodges, C. (2015). In search of quality: Using quality matters to analyze the quality of massive, open, online courses. *The International Review of Research in Open and Distributed Learning*, *16*(5), 83–101. doi:10.19173/irrodl.v16i5.2348

Luckmann, C. (1996). Defining Experiential Education. *Journal of Experiential Education*, *19*(1), 6–7. doi:10.1177/105382599601900101

Lynch, B., & Verdin, J. (1983). Job satisfaction in libraries. *The Library Quarterly*, *53*(4), 434–447. doi:10.1086/601406

Manz, C. C., & Sims, H. P. Jr. (1980). Self-management as a substitute for leadership: A social learning theory perspective. *Academy of Management Review*, *5*(3), 361–367. doi:10.5465/amr.1980.4288845

Margaryan, A., Bianco, M., & Littlejohn, A. (2015). Instructional quality of massive open online courses. *Computers and Education, 80*, 77–83. http://doi:10.1016/j.compedu.2014.08.005

Mariani, G., Carlesi, A. & Scarfò, A. (2018). Academic spinoffs as a value driver for intellectual capital. The case of the University of Pisa. *Journal of Intellectual Capital, 19*(1), 202-226.

Mawson, T. (2001). Ready! aim! inspire! Leadership in engineering. *Leadership and Management in Engineering*, *1*(2), 50–51. doi:10.1061/(ASCE)1532-6748(2001)1:2(50)

McGourty, J. (1999). *Alumni say communication skills are essential.* Retrieved March 3, 2003, from www.engineering.columba.edu/news/archive/engnews_s99/survey.html

McGrath, C. H., Hofman, J., Bajzikova, L., Harte, E., Lasakova, A., Pankowska, P., Sosso, S., Belanger, J., & Krivograd, J. (2016, June 31). *Governance and adaptation to innovative modes of higher education provision.* RAND Corporation. https://www.rand.org/pubs/research_reports/RR1571.html

McIsaac, M. S., & Gunawardena, C. N. I. (1996). Distance education. In D. H. Jonassen (Ed.), Handbook of Research for Educational Communications and Technology, (pp. 403-437). New York: Simon & Shuster Macmillan.

Medina, L. C. (2018). Blended learning: Deficits and prospects in higher education. *Australasian Journal of Educational Technology, 34*(1), 42-56. http://doi:10.14742/ajet.3100

Mehaffy, G. L. (2012). Challenge and change. *EDUCAUSE Review, 47*(5), 25–26. https://er.educause.edu/articles/2012/9/challenge-and-change

Meredith, J. (1992). Theory Building through Conceptual Methods. *International Journal of Operations & Production Management, 13*(5), 3–11. doi:10.1108/01443579310028120

Miao, Y., Sloep, P., Hummel, H., & Koper, R. (2009). Improving the unreliability of competence information: An argumentation to apply information fusion in learning networks. *International Journal of Continuing Engineering Education and Lifelong Learning, 19*(1), 1560–4624.

Miliam, J., Voorhees, R. A., & Bedard-Voorhees, A. (2004). Assessment of online education: Policies, practices and recommendations. *New Directions for Community Colleges, 126*(126), 73–85. doi:10.1002/cc.156

Miller, C. R. (1984). Genre as social action. *The Quarterly Journal of Speech, 70*(2), 151–167. doi:10.1080/00335638409383686

Mirfakhrai, M. K. (1991). Correlates of job satisfaction among academic librarians in the United States. *Journal of Library Administration, 14*(1), 117–131. doi:10.1300/J111v14n01_08

Moodle Analytics. (n.d.). Retrieved from https://docs.moodle.org/38/en/Analytics

Moodle Learning Analytics. (n.d.). Retrieved January 30, 2020, from https://docs.moodle.org/33/en/Learning_analytics

Mooney, C. (2008). IT on the campuses: What the future holds. *The Chronicle of Higher Education, 54*(30), 89.

Moore, M. G., & Kearsley, G. (2011). *Distance Education: A systems view of online learning*. Cengage Learning.

Morrison, J. L. (2007). Book Review: Northouse, Peter G. Leadership: Theory and Practice. *Journal of Education for Business, 83*(2), 116–117. doi:10.3200/JOEB.83.2.116-120

Mortensen, C. J., & Nicholson, A. M. (2015). The flipped classroom stimulates greater learning and is a modern 21st-century approach to teaching today's undergraduates. *Journal of Animal Science, 93*(7), 3722–3731. http://doi:10.2527/jas.2015-9087

Mor, Y., Ferguson, R., & Wasson, B. (2015). Editorial: Learning design, teacher inquiry into student learning and learning analytics: A call for action. *British Journal of Educational Technology, 46*(2), 221–229. doi:10.1111/bjet.12273

Mosley, A. L. (1998). A Behavioral Approach to Leadership: Implications for Diversity in Today's Organizations. *The Journal of Leadership Studies*, *5*(1), 39–50. doi:10.1177/107179199800500105

Mote, V. L. (1985). Development of Management Education in India. *Vikalpa*, *10*(4), 371–386. doi:10.1177/0256090919850402

Mo, Y., Dainty, A., & Price, A. (2006). The relevance of EQ to construction project management education and practice: an investigative framework. In D. Boyd (Ed.), *Proceedings of the 22nd Annual ARCOM Conference*, (Vol. 2, pp. 823–831). Association of Researchers in Construction Management.

Murugesan, V. (2019). Innovations in teaching. *Journal of Applied Science and Computations*, *6*(1), 2588–2596. https://www.researchgate.net/publication/331149966_Innovations_in_Teaching_Methods

Mykhailyshyn, H., Kondur, O., & Serman, L. (2018). Innovation of education and educational innovations in conditions of modern higher education institution. *Journal of Vasyl Stefanyk Precarpathian National University*, *5*(1), 9-16. http://doi:10.15330/jpnu.5.1.9-16

National Academy of Engineering (NAE). (2004). *The Engineer of 2020: Visions of Engineering in the New Century*. The National Academies Press.

Nat, M. (2015). A flipped classroom model for developing universities in developing countries. In *Global Learn* (pp. 597–604). Association for the Advancement of Computing in Education. https://www.researchgate.net/publication/275638404_A_Flipped_Classroom_Model_for_Developing_Universities_in_Developing_Countries

Nauta, M. M. (2007). Assessing college students' satisfaction with their academic majors. *Journal of Career Assessment*, *15*(4), 446–462. doi:10.1177/1069072707305762

Nelson, M. M., & Schunn, C. D. (2009). The nature of feedback: How different types of peer feedback affect writing performance. *Instructional Science*, *37*(4), 375–401. doi:10.100711251-008-9053-x

Norgard, R. T., Burkle, M., Tait, A., Guri-Rosenblit, S., & Canals, L. (2018). Universities of the future: Several perspectives on the future of higher education. In pushing the boundaries of higher education. *International Journal of Educational Technology in Higher Education*, *15*(46), 1–10. https://educationaltechnologyjournal.springeropen.com/universitiesofthefuture

O'Flaherty, J., & Phillips, C. (2015). The use of flipped classrooms in higher education: A scoping review. *The Internet and Higher Education*, *25*, 85–95. http://doi:10.1016/j.iheduc.2015.02.002

O'Grady, M., Barrett, G., Barrett, T., Delaney, Y., Hunt, N., Kador, T., & O'Brien, V. (2013). Reflecting on the need for problem triggers in multidisciplinary PBL. *AISHE-J: The All Ireland Journal of Teaching and Learning in Higher Education*, *5*(2), 1–21.

OECD. (2013). *The Survey of Adult Skills: Reader's Companion*. OECD Publishing., doi:10.1787/9789264204027-

Olivan, B., Masluk, B., Gascon, S., Fueyo Díaz, R., Aguilar-Latorre, A., Artola Magallon, I., & Magallón Botaya, R. (2019). The use of flipped classroom as an active learning approach improves academic performance in social work: A randomized trial in a university. *PloS One, 14*(4), 1-10. http://doi:10.1371/journal.pone.0214623

Olivé, D. M., Huynh, D. Q., Reynolds, M., Dougiamas, M., & Wiese, D. (2018). A supervised learning framework for learning management systems. *Proceedings of the First International Conference on Data Science, E-Learning and Information Systems - DATA 18*. 10.1145/3279996.3280014

Otero, I., Pantazatos, D., & Docampo, D. (2019). Good Practices Adopting Copyright for Creating and Reusing OERs in NGDLE Environment. *INTED2019 Proceedings*. doi: 10.21125/inted.2019.1359

Otero-González, I., Pantazatos, D., & Docampo, D. (2019). First Steps of Creating an Immersive Scenario for Improving Naval Skills. *ICERI2019 Proceedings*. doi: 10.21125/iceri.2019.0945

Owusu, G. M. Y., Essel-Anderson, A., Ossei Kwakye, T., Bekoe, R. A., & Ofori, C. G. (2018). Factors influencing career choice of tertiary students in Ghana: A comparison of science and business majors. *Education + Training*, *60*(9), 992–1008. doi:10.1108/ET-04-2017-0050

Paavola, S., Lipponen, L., & Hakkarainen, K. (2004). Models of innovative knowledge communities and three metaphors of learning. *Review of Educational Research*, *74*(4), 557–576. doi:10.3102/00346543074004557

Pantazatos, D., Marinos, C., Giotis, K., Vekris, A., Grammatikou, M., & Maglaris, V. (2018). SeLCont: Synchronized eLearning Content Toolkit. *INTED2018 Proceedings*. doi: 10.21125/inted.2018.2211

Paul, R., & Cowe, L. F. (2015). Engineering leadership education: A review of best practices. *Conference on Postsecondary Learning and Teaching*. University of Calgary. 10.18260/p.23972

Payne, A., Frow, P., & Eggert, A. (2017). The customer value proposition: Evolution, development, and application in marketing. *Journal of the Academy of Marketing Science*, *45*(4), 467–489. doi:10.100711747-017-0523-z

Pearce, N., Weller, M., Scanlon, E., & Kinsley, S. (2010). Digital scholarship considered: How new technologies could transform academic work. *In Education, 16*(1), 33-44. https://ineducation.ca/ineducation/article/view/44/508

Peterson, A. (2018). *Understanding innovative pedagogies: Key themes to analyze new approaches to teaching and learning*. http://www.oecd.org/officialdocuments/publicdisplaydocumentpdf/?cote=EDU/WKP(2018)8&docLanguage=En

Peter, Y., & Vantroys, T. (2005). Platform support for pedagogical scenarios. *Journal of Educational Technology & Society*, *8*(3), 122–137.

Piccoli, G., Ahmad, R., & Ives, B. (2001). Web-based virtual learning environments: A research framework and a preliminary assessment of effectiveness in basic IT skills training. *Management Information Systems Quarterly*, *25*(4), 401–426. doi:10.2307/3250989

Pilli, O., & Admiraal, W. (2016). A taxonomy of massive open online courses. *Contemporary Educational Technology, 7*(3), 223-240. https://www.researchgate.net/publication/305602116_A_Taxonomy_of_Massive_Open_Online_Courses

Pinelli, T. E., Barclay, R. O., Keene, M. L., Kennedy, J. M., & Hecht, L. F. (1995). From student to entry-level professional: Examining the role of language and written communications in the reacculturation of aerospace engineering students. *Technical Communication (Washington), 42*, 492–507.

Pisanu, F. (2014). Educational innovation and technology: A need for integration. *Perspectives of Innovations, Economics and Business, 14*(2), 103-108. http://doi:10.15208/pieb.2014.12

Pisoni, G. (2019). Strategies for pan-European implementation of blended learning for innovation and entrepreneurship education. *Education Sciences, 9*(2), 124-137. http://doi:10.3390/educsci9020124

Pitts, S., Klosterman, S., & McGonagle, S. (2013). A successful approach to educating engineering leaders at the graduate level. *Proceedings of the Canadian Engineering Education Association (CEEA)*. 10.24908/pceea.v0i0.4819

Ragin, C. C. (1994). *Constructing social research.* Pine Forge Press.

Raiden, A. B., Dainty, A., & Neale, R. H. (2004). Current barriers and possible solutions to effective project team formation and deployment within a large construction organization. *International Journal of Project Management, 22*(4), 309–316. doi:10.1016/j.ijproman.2003.08.002

Rasoal, C., Danielsson, H., & Jungert, T. (2012). Empathy among students in engineering programmes. *European Journal of Engineering Education, 37*(5), 427–435. doi:10.1080/03043797.2012.708720

Reave, L. (2004). Technical communication instruction in engineering schools: A survey of top-ranked US and Canadian programs. *Journal of Business and Technical Communication, 18*(4), 452–490. doi:10.1177/1050651904267068

Reichheld, F. (1996). Learning from Customer Defections. *Harvard Business Review, 29*(6), 755–756.

Rennstam, J., & Kärreman, D. (2019). Understanding control in communities of practice: Constructive disobedience in a high-tech firm. *Human Relations, 73*(6), 864–890. doi:10.1177/0018726719843588

Rhode, J. (2009). Interaction Equivalency in Self-Paced Online Learning Environments: An Exploration of Learner Preferences. *The International Review of Research in Open and Distributed Learning, 10*(1). Advance online publication. doi:10.19173/irrodl.v10i1.603

Riggio, R. E. (2008). Leadership development: The current state and future expectations. *Consulting Psychology Journal: Practice and Research, 60*(4), 383–392. doi:10.1037/1065-9293.60.4.383

Riley, L. A., Furth, P. M., & Zellmer, J. T. (2000). *Assessing our engineering alumni: Determinants of success in the workplace*. Retrieved November 12, 2003, from www.ece.nmsu.edu/pfurth/papers/00/ASEEABET.pdf

Roehling, P. V., Luna, L. M., Richie, F. J., & Shaughnessy, J. J. (2017). The benefits, drawbacks, and challenges of using the flipped classroom in an introduction to psychology course. *Teaching of Psychology, 44*(3), 183–192. http://doi:10.1177/0098628317711282

Rosser, S. V. (2004). *The science glass ceiling: Academic women scientists and the struggle to succeed*. Routledge.

Rovai, A. P. (2002a). Building sense of community at a distance. *International Review of Research in Open and Distance Learning, 3*(1). Online at: http://www.irrodl.org/index.php/irrodl/article/view/79/152

Rovai, A. (2002). Building Sense of Community at a Distance. *International Review of Research in Open and Distance Learning*, *3*(1). Advance online publication. doi:10.19173/irrodl.v3i1.79

Sageev, P., & Romanowski, C. J. (2001). A message from recent engineering graduates in the marketplace: Results of a survey on technical communication skills. *Journal of Engineering Education*, *90*(4), 685–697. doi:10.1002/j.2168-9830.2001.tb00660.x

Salovey, P., & Mayer, J. (1990). Emotional intelligence. *Imagination, Cognition and Personality*, *9*(3), 185–211. doi:10.2190/DUGG-P24E-52WK-6CDG

Samadi, A. (2016). The Impact of Intellectual Capital on Students' Creativity from Teachers'. *International Journal of Humanities and Cultural Studies*, *2*(4), 1584–1591.

Sankar, C. S., & Raju, P. K. (2011). Use of presage-pedagogy-process-product model to assess the effectiveness of case study methodology in achieving learning outcomes. *Journal of STEM Education: Innovations & Research*, *12*(7), 45–56.

Sansone, N., Cesareni, D., Ligorio, M. B., Bortolotti, I., & Buglass, S. L. (2019). Developing knowledge work skills in a university course. Research Papers in Education. Doi:10.1080/02671522.2019.1677754

Sansone, N., Bortolotti, I., & Buglass, S. (2016). The Trialogical Learning Approach in Practice(s): Reflections from Pedagogical Cases. *Qwerty, Open and Interdisciplinary Journal of Technology, Culture and Education, 11(1).*

Sansone, N., Cesareni, D., Bortolotti, I., & Buglass, S. (2019). Teaching technology-mediated collaborative learning for trainee teachers. *Technology, Pedagogy and Education*, *28*(3), 381–394. doi:10.1080/1475939X.2019.1623070

Savin-Baden, M. (2007). *A practical guide to problem-based learning online*. Routledge. doi:10.4324/9780203938140

Scardamalia, M., & Bereiter, C. (2006). Knowledge Building: Theory, pedagogy, and technology. In K. Sawyer (Ed.), *Cambridge Handbook of the Learning Sciences* (pp. 97–118). Cambridge University Press.

Scardamalia, M., Bransford, J., Kozma, B., & Quellmalz, E. (2012). New assessments and environments for knowledge building. In P. Griffin, B. McGaw, & E. Care (Eds.), *Assessment and teaching of 21st century skills* (pp. 231–300). Springer. doi:10.1007/978-94-007-2324-5_5

Schaeffer, L. D. (2002). The Leadership Journey. *Harvard Business Review*, 42–47. PMID:12389460

Schneider, D. K., Synteta, P., Frété, C., Girardin, S., & Morand, S. (2003). Conception and implementation of rich pedagogical scenarios through collaborative portal sites: clear focus and fuzzy edges. In *Proceedings of the International Conference on Open & Online Learning, ICOOL, International Conference on Open and Online Learning*. University of Mauritius.

Schuck, S., & Kearney, M. (2008). Classroom-based use of two educational technologies: A sociocultural perspective. *Contemporary Issues in Technology and Teacher Education: Current Practices, 8*, 394–406. https://citejournal.org/volume-8/issue-4-08/current-practice/classroom-based-use-of-two-educational-technologies-a-sociocultural-perspective/

Sciarrone, F. (2018). Machine Learning and Learning Analytics: Integrating Data with Learning. *17th International Conference on Information Technology Based Higher Education and Training (ITHET)*. 10.1109/ITHET.2018.8424780

Scott, G., Coates, H., & Anderson, M. (2008). *Learning leaders in times of change: Academic leadership capabilities for Australian higher education*. University of Western Sydney and Australian Council for Educational Research.

Seat, E., Parsons, J. R., & Poppen, W. A. (2001). Enabling engineering performance skills: A program to teach communication, leadership, and teamwork. *Journal of Engineering Education, 90*(1), 1–12. doi:10.1002/j.2168-9830.2001.tb00561.x

Secundo, G., Massaro, M., Bagnoli, C., & Dumay, J. (2018). Intellectual capital management in the fourth stage of IC research: A critical case study in University settings. *Journal of Intellectual Capital, 19*(1), 157–177. doi:10.1108/JIC-11-2016-0113

Seemiller, C., & Murray, T. (2013). The common language of leadership. *The Journal of Leadership Studies, 7*(1), 33–45. doi:10.1002/jls.21277

Seibert, S. E., Kraimer, M. L., & Crant, J. M. (2001). What do proactive people do? A longitudinal model linking proactive personality and career success. *Personnel Psychology, 54*(4), 845–874. doi:10.1111/j.1744-6570.2001.tb00234.x

Sellgren, S., Ekvall, G., & Tomson, G. (2006). Leadership styles in nursing management: Preferred and perceived. *Journal of Nursing Management, 14*(5), 348–355. doi:10.1111/j.1365-2934.2006.00624.x PMID:16787469

Sergis, S., & Sampson, D. G. (2017). Teaching and Learning Analytics to Support Teacher Inquiry: A Systematic Literature Review. Learning Analytics: Fundaments, Applications, and Trends Studies in Systems, Decision and Control, 25–63. doi:10.1007/978-3-319-52977-6_2

Seymour, E., Hunter, A.-B., Laursen, S. L., & Deantoni, T. (2004). Establishing the Benefitsof Research Experiencesfor Undergraduates in the Sciences: First Findings from a Three-Year Study. *Science Education*, *88*(4), 493–534. doi:10.1002ce.10131

Sfard, A. (1998). On two metaphors for learning and the dangers of choosing just one. *Educational Researcher*, *27*(2), 4–13. doi:10.3102/0013189X027002004

Shivers-Blackwell, S. L. (2004). Reactions to outdoor teambuilding initiatives in MBA education. *Journal of Management Development*, *23*(7), 614–630. doi:10.1108/02621710410546632

Shute, V. J. (2008). Focus on formative feedback. *Review of Educational Research*, *78*(1), 153–189. doi:10.3102/0034654307313795

Shyu, J. F. (1985). A study on job satisfaction of the university library staff. *Journal of Library and Information Science*, *11*, 207–234.

Sibthorp, J., Paisley, K., & Gookin, J. (2007). Exploring Participant Development Through Adventure-Based Programming: A Model from the National Outdoor Leadership School. *Leisure Sciences*, *29*(1), 1–8. doi:10.1080/01490400600851346

Simonson, S. R. (2017). To flip or not to flip: What are the questions? *Education Sciences, 7*(3), 71-75. http://doi:10.3390/educsci7030071

Slavin, R. E. (2011). Instruction Based on Cooperative Learning. In R. E. Mayer & P. A. Alexander (Eds.), *Handbook of Research on Learning and Instruction* (pp. 344–360). Taylor & Francis.

Smithers, G. L., & Walker, D. H. T. (2000). The effect of the workplace on motivation and demotivation of construction professionals. *Construction Management and Economics*, *18*(7), 833–841. doi:10.1080/014461900433113

Smolin, L. I., & Lawless, K. A. (2003). Becoming literate in the technological age: New responsibilities and tools for teachers. *The Reading Teacher*, *56*(6), 570–577.

Smyth, H. (2000). *Marketing and Selling Construction Services*. Blackwell.

Sougui, A. O., Bon, A. T., & Hassan, H. M. (2015). The Impact of Leadership Styles on Employees' performance in Telecom Engineering companies. *Australian Journal of Basic and Applied Sciences*, *8*(4).

Spector, J. M., Merrill, M. D., Merrienboer, J. V., & Driscoll, M. P. (2008). *Handbook of research on educational communications and technology* (3rd ed.). Lawrence Erlbaum Associates.

Stajkovic, A. D., & Luthans, F. (1997). A meta-analysis of the effects of organizational behavior modification on task performance, 1975-95. *Academy of Management Journal*, *40*(5), 1122–1149.

Stewart, G. L. (2010). The past twenty years: Teams research is alive and well at the journal of management. *Journal of Management, 36*(4), 801–805. doi:10.1177/0149206310371512

Stohr, C., & Adawi, T. (2018). Flipped classroom research: From "black box" to "white box" evaluation. *Educational Sciences, 8*(1), 22-25. http://doi:10.3390/educsci8010022

Storey, J. e. (2016). *Leadership in Organizations: Current issues and key trends* (3rd ed.). Routledge. doi:10.4324/9781315695792

Strijbos, J. W., Martens, R., Jochems, W. M. G., & Broers, N. J. (2004). The effect of functional roles on group efficiency: Using multilevel modeling and content analysis to investigate computer-supported collaboration in small groups. *Small Group Research, 35*(2), 195–229. doi:10.1177/1046496403260843

Sugahara, S., & Boland, G. (2009). The Accounting Profession as a Career Choice for Tertiary Business Students in Japan-A Factor Analysis. *Accounting Education, 18*(3), 255–272. doi:10.1080/09639280701820035

Sultanova, G., Svyatov, S., & Ussenbayev, N. (2018). Transmitting competencies at universities in Kazakhstan: Intellectual capital of teachers. *Journal of Intellectual Capital, 19*(1), 112–134. doi:10.1108/JIC-04-2017-0058

Summers, M., Davis, B., & Tomovic, C. (2004). When engineering and technology skills are not enough: Engineering leaders out of their element. *Proceedings of the CIEC Conference.*

Sundstrom, E., McIntyre, M., Halfhill, T., & Richards, H. (2000). Work groups: From the hawthorne studies to work teams of the 1990s and beyond. *Group Dynamics, 4*(1), 44–67. doi:10.1037/1089-2699.4.1.44

Sveiby, K. E. (2010). *Methods for measuring intangible assets*. Available at: www.sveiby.com/portals/0/articles/IntangibleMethods.htm

Szadziewska, A., & Kujawski, J. (2017). Advantages and disadvantages of the blended-learning method used in the educational process at the faculty of management at the university of Gdansk, in the opinion of undergraduate students. In *10th annual international conference of education, research and innovation* (vol. 2017, pp. 3938-3946). http://doi:10.21125/iceri.2017.1051

Tham, C. M., & Werner, J. M. (2005). Designing and evaluating e-learning in higher education: A review and recommendations. *Journal of Leadership & Organizational Studies, 11*(2), 15–25. doi:10.1177/107179190501100203

The Boeing Company. (n.d.). http://www.boeing.com

Thompson, A. D., Grahek, M., Phillips, R. E., & Fay, C. L. (2008). The Search for Worthy Leadership. *Consulting Psychology Journal: Practice and Research, 60*(4), 366–382. doi:10.1037/1065-9293.60.4.366

Thornton, J. K. (2001). African American female librarians: A study of job satisfaction. *Journal of Library Administration, 33*(1/2), 141–164. doi:10.1300/J111v33n01_10

Thurnell, D. (2004). Emotional intelligence and the value management facilitator. In F. Khosrowshahi (Ed.), *Proceedings of the 20th Annual ARCOM Conference*, (Vol. 2, pp. 918–922). Association of Researchers in Construction Management.

Timmis, S., Broadfoot, P., Sutherland, R., & Oldfield, A. (2015). Rethinking assessment in a digital age: Opportunities, challenges, and risks. *British Educational Research Journal, 42*(3), 454–476. http://doi:10.1002/berj.3215

Tonso, K. L. (2006). Teams that work: Campus culture, engineer identity, and social interactions. *Journal of Engineering Education*, *95*(1), 25–37. doi:10.1002/j.2168-9830.2006.tb00875.x

Trentin, G. (2003). E-learning come sistema complesso [E-learning as a complex system]. *IJET*, *11*(3), 47–52.

Trottier, T., Wart, M. V., & Wang, X. (2008). Examining the Nature and Significance of Leadership in Government Organizations. *Public Administration Review*, *68*(2), 319–333. doi:10.1111/j.1540-6210.2007.00865.x

Tryggvason, G., Thouless, M., Dutta, D., Ceccio, S. L., & Tilbury, D. M. (2001). The new mechanical engineering curriculum at the University of Michigan. *Journal of Engineering Education*, *90*(3), 437–448. doi:10.1002/j.2168-9830.2001.tb00624.x

UNESCO. (2010). *Engineering: Issues Challenges and Opportunities for Development*. UNESCO Publishing.

Uskov, V., Bakken, J. P., Aluri, L., Rachakonda, R., Rayala, N., & Uskova, M. (2018). Smart pedagogy: Innovative teaching and learning strategies in engineering education. *World Engineering Education Conference, 2018*(2018), 1-6. http://doi:10.1109/edunine.2018.8450962

Van Zyl, C., & de Villiers, C. (2011). Why some students choose to become chartered accountants (and others do not). *Meditari Accountancy Research*, *19*(1/2), 56–74. doi:10.1108/10222521111178637

VDE. (2003). *Young Professionals*. Available online at: www.vde.com/vde_en/

Virakul, B., & McLean, G. N. (2012). Leadership development in selected leading Thai companies. *The Journal of Leadership Studies*, *6*(1), 6–22. doi:10.1002/jls.21224

Vohra, N., Rathi, N., & Bhatnagar, D. (2015). Developing Leadership Skills among EMBA Students: Innovations in Design. *The Journal for Decision Makers*, *40*(1), 15–27. doi:10.1177/0256090915573612

Vrasidas, C. (2000). Constructivism Versus Objectivism: Implications for Interaction, Course Deign, and Evaluation in Distance Education. *International Journal of Educational Telecommunications*, *6*(4), 339–362.

Vroom, V. H., & Jago, A. G. (2007). The Role of the Situation in Leadership. *The American Psychologist*, *62*(1), 17–24. doi:10.1037/0003-066X.62.1.17 PMID:17209676

Vygotsky, L. (1978). *Mind in Society* (M. Cole, Trans.). Harvard University Press.

Vygotsky, L. S. (1978). *Mind in society*. Harvard University Press.

Wagner, E. D. (1994). In support of a functional definition of interaction. *American Journal of Distance Education*, *8*(2), 6–29. doi:10.1080/08923649409526852

Walker, M. (2001). Engineering identities. *British Journal of Sociology of Education*, *22*(1), 75–89. doi:10.1080/01425690020030792

Walther, J., Miller, S. E., & Sochacka, N. W. (2017). A model of empathy in engineering as a core skill, practice orientation, and professional way of being. *Journal of Engineering Education*, *106*(1), 123–148. doi:10.1002/jee.20159

Wang, K., & Zhu, C. (2019). MOOC-based flipped learning in higher education: Students' participation, experience, and learning performance. *International Journal of Educational Technology in Higher Education*, *16*(1), 33-40. http://doi:10.118641239-019-0163-0

Wang, H., & Guan, B. (2018). The Positive Effect of Authoritarian Leadership on Employee Performance: The Moderating Role of Power Distance. *Frontiers in Psychology*, *9*, 357. doi:10.3389/fpsyg.2018.00357 PMID:29628902

Wartfoskij, M. (1973). *Models*. D. Reidel.

Watson, S., & Vasilieva, E. (2007). Wilderness thinking: A novel approach to leadership development. *Industrial and Commercial Training*, *39*(5), 242–245. doi:10.1108/00197850710761927

Watzlawick, P., Beavin, J. H., & Jackson, D. D. (1971). *Pragmatica della comunicazione umana*. Astrolabio.

Webb, N. (1985). *Student interaction and learning in small groups: A research summary*. Learning to Cooperate, Cooperating to Learn. doi:10.1007/978-1-4899-3650-9_6

Weinberger, A. (2008). *CSCL scripts: Effects of social and epistemic scripts on computer-supported collaborative learning*. VDM.

Wenger, E., Trayner, B., & De Laat, M. (2011). *Promoting and assessing value creation in communities and networks: A conceptual framework*. Ruud de Moor Centrum. Available at http://www.knowledgearchitecture.com/downloads/Wenger_Trayner_DeLaat_Value_creation.pdf

What is xAPI aka the Experience API or Tin Can API. (2020). Retrieved 30 January 2020, from https://xapi.com/overview

Whetten, D. A., & Cameron, K. S. (2002). *Developing Management Skills* (5th ed.). Prentice Hall.

Whitaker, B., & Arensdorf, J. (2018). Big Question for Leadership Education. *The Journal of Leadership Studies*, *12*(2), 46–48. doi:10.1002/jls.21575

Whiteley, T. R. (2006). Using the Socratic Method and Bloom's Taxonomy of the Cognitive Domain to Enhance Online Discussion, Critical Thinking and Student Learning. *Developments in Business Simulation and Experiential Learning*, *33*, 65–70.

Williams, J. M. (2001). Transformations in Technical Communication Pedagogy: Engineering, Writing, and the ABET Engineering Criteria 2000. *Technical Communication Quarterly*, *10*(2), 149–167. doi:10.120715427625tcq1002_3

Williams, S. D., Graham, T. S., & Baker, B. (2003). Evaluating outdoor experiential training for leadership and team building. *Journal of Management Development*, *22*(1), 45–59. doi:10.1108/02621710310454851

Wilson, K., Ruch, G., Lymbery, M., & Cooper, A. (2008). *Social work: An introduction to contemporary practice*. Pearson.

Xu, M., David, J. M., & Kim, S. H. (2018). The Fourth Industrial Revolution: Opportunities and Challenges. *International Journal of Financial Research*, *9*(2), 90. doi:10.5430/ijfr.v9n2p90

Yadav, A., & Patwardhan, A. (2016). Use and impact of web 2.0 tools in higher education: A literature review. In S. Parmar & A. K. Siwach (Eds.), *Academic libraries in electronic environment* (pp. 218-246). Intellectual Foundation. http://doi:10.13140/RG.2.1.2748.6965/1

Yurtseven, H. O. (2002). How does the image of engineering affect student recruitment and retention? A perspective from the USA. *Global Journal of Engineering Education*, *6*(1), 17–23.

About the Contributors

Despo Ktoridou is a professor at the University of Nicosia in Cyprus. She is also the Head of the Management & MIS Department of the School of Business since 2012. Despo is a Certified Engineer of the Cyprus Scientific and Technical Chamber (ETEK) and a member of IEEE and ACM Associations. In the period 1993-1999 Despo worked as a Senior Computer Engineer for various organizations in Cyprus. In 1999 she entered academia, as an Assistant Professor of Educational Technologies until 2007. In 2007, she was promoted to Associate Professor and in 2019, she was ranked as Professor. Her research focuses on the areas of ICT Integration in the Educational Practice; Technology and Innovation Management Pedagogies; and Open and Online Distance Learning Practices. She has presented and published papers, chaired Sessions, served as reviewer and member of the Program Committee for numerous refereed international conferences. She has contributed several chapters in books, served as a guest editor and reviewer and published several papers in refereed journals. She has also participated in EU and local funded programs and has been invited by foreign universities as a guest lecturer. Professor Ktoridou completed her BSc. and MSc. of Science in Engineering, in the field of Computer Systems & Networks in the Engineering & Science Department of Saint Petersburg Electrotechnical University in Russia. Her PhD thesis title was 'Expert Systems in Education - Distance Learning Educational System for Early Childhood Education: Methods, Facilities and Means'.

* * *

Zeyad Abualhamael admires working in a professional environment. Always, seeking challenging and rewarding roles which will fully initialize his talents and his capability as a good communicator, with excellent English and Arabic language, time management and leadership skills with the ability to multi task unrelated activities. Zeyad appreciates working under pressure and makes an effective contribution to any situation. Within the education sector, Zeyad is currently a fellow lecturer at Manchester Metropolitan University in Manchester. Holding an external examiner position in Bedford University in London and MSA Cairo in Egypt. Within the sports sector, Zeyad is a consultant on the board of the Talent discovery and development within the Saudi Arabian football federation. Within the food sector, Zeyad also is a consultant for several business, cafes, restaurants and fast-food chains. Finally, Zeyad is the head and director of Project Square One – Leadership in Practice.

Anna-Maria Andreou has a Master's degree (M.A. in Descriptive and Applied Linguistics, 1996, University of Essex, UK) and a PhD (P.h.D in Second Language Acquisition and Education, 2014, IOE, University of London, England). She was a lecturer of English Language and Linguistics in higher education for 21 years and currently works for the Cyprus Ministry of Education, Culture, Sport and Youth. She participated in European programs (Tempus projects) for 5 years and has presented her research work at international scientific conferences. Her research interests include input-based approaches (i.e. structured-input), production approaches (e.g. text manipulation, text-creation) and corrective feedback approaches (e.g. recasts, requests for clarification) to teaching grammar, as well their implications for SLA theory development, language pedagogy and language policy, and materials development and analysis.

George Andreakos is an Electrical Engineer working at Electricity North West (UK). He is a graduate of the University of Manchester, gaining a BA in Electrical Engineering and Electronics and a Masters Degree from the Manchester Metropolitan University in Comuter and Network Technology. He specialises in all Electrical aspects of Building Services Engineering, while he manages and supervises engineering projects across the North West of England.

Marilena Antoniadou is a Reader in Management and Business Psychology. She specialises in the role of emotions and emotional management in the workplace. She is the holder of the Newer Researchers Award for 2015 by the Society of Research into Higher Education. She is also a Senior Fellow of the Higher Education Academy, a Chartered Academic member of the CIPD (MCIPD), a Certified Management and Business Educator (CMBE) and a committee member of the Association for Business Psychology. Her research activities pertain to organisational phenomena that involve emotional processes, such as emotional intelligence, leadership and values. She has published within the fields of educational management and organisational psychology

Donatella Cesareni (Ph.D.) is Associated Professor in Educational Studies in the faculty of Medicine and Psychology, University of Rome " Sapienza". She is a member of the teacher staff of the PHD in "Psychology of interaction, communication and socialization processes" and since 2006 is the co-editor of the scientific review "QWERTY, journal of technology, culture and education". Her research interests are focused on collaborative use of technologies in education. She has participated in several research programs, concerning the use of technologies at schools and universities, both at international and national level.

Mark Crowder is a Senior Lecturer and Education Lead for the Department of Strategy, Enterprise and Sustainability at Manchester Metropolitan University in the UK. He is also a senior lecturer in Strategy and Business Psychology. He has more than 20 years' management, and studied at Liverpool John Moores University and the University of Liverpool before gaining his PhD in cognitive psychology at the University of Chester. He is a Senior Fellow of the Higher Education Academy and a Fellow of the Chartered Management Institute. Mark's research interests are split between educational management and cognitive psychology.

Leonidas Efthymiou joined the University of Nicosia as full-time lecturer in September 2018, having worked previously at Unicaf University, Intercollege Larnaca and the University of Leicester. He holds a PhD and MSc in Management from the University of Leicester, a degree (BA Hons) in Hospitality Management from the University of Derby and a Diploma in Hotel & Catering Management from the Higher Hotel Institute Cyprus. His research interests include a wide range of business phenomena, varying from emotional, affective and aesthetic labour, to cashless payments, corporate legitimacy and sociology of finance. He has co-edited the book, titled, "The Book of Payments: Historical and Contemporary Views on Retail Monetary Exchange" with Prof. Bernardo Batiz-Lazo (Palgrave Macmillan). His PhD Thesis received the 2012 Best Dissertation award at the Academy of Management Meeting, held at Boston, Massachusetts. Research available at: https://independent.academia.edu/LeonidasEfthymiou. He currently acts as the coordinator of the BBA Hospitality Management program.

Epaminondas Epaminonda is an Assistant Professor in the Management Department of the Business School of the University of Nicosia. He is also the Associate Head of the Department and the Coordinator of the Bachelor Program in Business Administration. He has 18 years of university experience, having taught a variety of courses primarily in the areas of Management, Organisational Behaviour and Human Resource Management. He also has seven years of industry work experience, having worked in the banking, tourism and market research sectors. His research interests are in the areas of comparative management, values change, job satisfaction and business-society interactions. Dr. Epaminonda holds BA and MA degrees from the University of Cambridge, an MBA from Imperial College and a PhD from Manchester Business School.

Maty Grammatikou is a Senior Researcher in the Network Management and Optimal Design Laboratory (NETMODE) in NTUA, Greece. Her research interests include topics on management and Optimal Design of Computer Communication Networks, Quality of Service, Education and Learning Systems, Security and Intrusion Detection Systems also Cloud Computing Architectures and web-based technologies. She has an extensive experience in research projects related to digital technologies and education. She has a PhD in Computer Science from the NTUA of Greece. She has several publications in conferences, books and journals related to her research interests. She is a member of IEEE and TEE and also she is a member of the WG7/TC48 Working Group and TM and DMT Forums.

Marcos Komodromos is an Assistant Professor of Communications & Public Relations at the University of Nicosia and a Chartered PR Practitioner (London, UK). He has earned two Masters Degrees in Communications and in Management and completed his PhD in organizational justice, trust and the management of change. Currently, he is a consultant in strategic communication, digital media and marketing communication in organizations in Cyprus and abroad, and he is a certified trainer and coach for employees in the areas of communication and public relations, marketing, social media, leadership, media training, and management. Dr Komodromos has actively worked in the media industry for the past 15 years and until now he has delivered more than 10,000 hours of training to working professionals. Marcos Komodromos is a Member of the Chartered Institute of Public Relations (CIPR) of Great Britain, a Chartered PR practitioner (Ch. PR) and has been an active member of many Professional Associations and organizations in Cyprus and the U.K.

Vasilis Maglaris holds an Engineering Degree from the National Technical University of Athens – NTUA (Athens, Greece, 1974) and a Ph.D. degree from Columbia University (New York, USA, 1979). Between 1979 and 1989 he held industrial and academic positions in the USA, all in advanced electronic communications. From 1989 to 2019 he was with the faculty of the School of Electrical & Computer Engineering at NTUA teaching and performing research on Internet technologies, directing the Network Management & Optimal Design (NETMODE) Laboratory that he established. He was responsible for the development of the NTUA Campus LAN and of GRNET, the National Research & Education Network (NREN) of Greece. From 2004 to 2012 he was the GÉANT Policy Committee Chairman, the governance body of the advanced Internet serving the 37 NRENs of the extended European Research Area. From July 2012 to June 2013 he served as General Secretary for Research & Technology appointed by the Greek coalition Government, on hold from his duties at NTUA. In 2020 the NTUA Senate conferred upon him the title of Professor Emeritus, enabling him to continue his teaching and research activities beyond his retirement in 2019.

Venugopal Merugu is an ad-hoc faculty in the School of Management, National Institute of Technology (NIT), Warangal. He received his Ph.D and Masters in Finance and Marketing from the School of Management, NIT Warangal. The major research areas include corporate finance, financial management and general management.

Yianna Orphanidou is the Associate Head of the Hospitality,Tourism and Sports Management Department of the of the University of Nicosia .She held the position of chairperson of the Institute of Hospitality (IH), Cyprus Branch (former HCIMA). She is a Fellow member of IH; and a board member of the Cyprus Sustainable Tourism Initiative (CSTI). She holds an HND in Hotel Management (F&B specialization) from HHIC, a BA (HONS) in Hospitality Management and an MSc in Hospitality and Tourism Education from Surrey University (UK). She is an accredited trainer from the Human Resource Department Authority of Cyprus . Her research interests include the design and development of Tourism/hospitality management competency model, the design of curriculum and interactive methods of teaching hospitality and tourism modules. She has participated in the technical teams of various EU funded projects with the last decade focusing in ECVET projects. She has more than 20 years of Hospitality/Tourism academic and professional experience.

Dimitris Pantazatos is a research associate and a PhD Candidate at Network and Optimal Design Laboratory (NETMODE), School of Electrical and Computer Engineering (ECE), National Technical University of Athens (NTUA). His main research interests and related efforts/publications are focused on video streaming technologies and synchronization of multimedia content, AR and VR applications in Education, as well as, Multi-Agent Environments and Applications on Education.

Nadia Sansone is a Researcher in Experimental Pedagogy and Technology Enhanced Learning at Unitelma Sapienza University of Rome where she is the eL-earning Responsible and Head of teachers and tutors' Training. Her main research interests concern models for the design and analysis of e-learning paths supported by Learning Management Systems in which learners experience active approaches promoting theoretical and practical learning. Author of several international publications, including communications to conferences, book's chapters and journal articles. She is Chief-editor of QWERTY - Open and Interdisciplinary Journal of Technology, Culture and Education.

Rahul Thangeda is an assistant professor in the School of Management, National Institute of Technology (NIT), Warangal. He received his Ph.D and Masters in Marketing and Human Resource from the School of Management, NIT Warangal. The major research areas include marketing management, business analytics and general management.

Alex Zarifis has taught and carried out research at several universities including the University of Manchester, the University of Liverpool and the University of Mannheim. He also participated in creating and delivering the first university module on blockchain technologies in the world. He has worked on large EU and UK funded research projects at the Karlsruhe Institute of Technology (KIT) and Loughborough University. He obtained his PhD from the University of Manchester. His research interests include e-business, the sharing economy, artificial intelligence, trust, information privacy, blockchain and Fintech. Dr Alex has published extensively on trust in e-commerce. His research has explored the role of trust in various aspects of e-business including: The level of trust needed for different business models in the sharing economy, the impact of ransomware attacks on trust, how trust develops in online collaboration, online collaborative learning and multi-channel retail.

Index

www.ingramcontent.com/pod-product-compliance
Lightning Source LLC
LaVergne TN
LVHW081321300926
849161LV00008B/1843